Building Sustainability in East Asia

Building Sustainability in East Asia

Policy, Design, and People

Vincent S Cheng

Jimmy C Tong

WILEY Blackwell

This edition first published 2017
© 2017 by John Wiley & Sons Ltd

Registered Office
John Wiley & Sons Ltd, The Atrium, Southern Gate, Chichester, West Sussex, PO19 8SQ, United Kingdom.

Editorial Offices
9600 Garsington Road, Oxford, OX4 2DQ, United Kingdom.
The Atrium, Southern Gate, Chichester, West Sussex, PO19 8SQ, United Kingdom.

For details of our global editorial offices, for customer services and for information about how to apply for permission to reuse the copyright material in this book please see our website at www.wiley.com/wiley-blackwell.

Library of Congress Cataloging-in-Publication Data

Names: Cheng, Vincent S., author. | Tong, Jimmy C., author.
Title: Building sustainability in East Asia : policy, design, and people / Vincent S. Cheng, Jimmy C. Tong.
Description: Chichester, West Sussex : John Wiley & Sons, 2017. | Includes bibliographical references and index.
Identifiers: LCCN 2016051176 (print) | LCCN 2017010814 (ebook) | ISBN 9781119277002 (cloth) |
    ISBN 9781119276982 (pdf) | ISBN 9781119277026 (epub)
Subjects: LCSH: Sustainable construction–East Asia. | Sustainable buildings–East Asia. |
    Sustainable urban development–East Asia.
Classification: LCC TH880 .C464 2017 (print) | LCC TH880 (ebook) | DDC 338.95/07–dc23
LC record available at https://lccn.loc.gov/2016051176

Cover image: Andy Yeung Photography
Cover design by Wiley

Set in 10/12.5pt Minion by SPi Global, Pondicherry, India

10 9 8 7 6 5 4 3 2 1

Contents

Section 5   Way forward   **235**

About the authors

Dr. Vincent S Cheng is the Director of Building Sustainability Group of Arup at East Asia. He has professional experience in building energy efficiency and sustainable building environment design and government consultation studies, including Taiwan EPA's Low Zero Carbon Pilot Project, CEPAS of Buildings Department, Life Cycle Assessment Tool and Air Ventilation Assessment. Dr. Cheng also specializes in sustainable master planning, low/zero carbon design, and LEED & BEAM Plus certification. He is a director of Hong Kong Green Building Council and also a member in the advisory committee of School of Energy and Environment, the City University of Hong Kong.

Dr. Jimmy C Tong is East Asia Energy Skill Leader and an Associate at Arup at East Asia focusing on Building Sustainability. A recognised industry leader within the energy business, Dr. Tong has applied his expertise in energy systems in various sectors, including wind and renewable energy, infrastructure and building services, and product and system development in the manufacturing of electronics, ventilation equipment, and filtration equipment. He obtained a PhD specialising in computational fluid flow and heat transfer from the University of Minnesota. He is also a guest lecturer at several universities in Hong Kong on the subject of energy and sustainability and co-authored book publication and book chapters and publications in archival, refereed journals.

Foreword

The current pace and scale of urbanisation is unprecedented in human history, presenting critical environmental and resource constraint challenges. This demands urgent solutions and actions that lead to a low carbon and sustainable outcome.

We need to make cities better in the developed economies, by retrofitting buildings and infrastructure, and we need to make better new cities in the developing economies. Many cities in East Asia have been undergoing rapid development in recent years whilst trying to meet the challenges. The experience plays an important role in global urban transformation as, if done right, sustainable Asian cities can serve as models for other emerging cities around the world.

The authors have for many years been researching the issues involved, developing tools and solutions, and designing many projects of different scales at different locations. In this book, they combine their practice of advanced engineering techniques with their passion to shape a greener and better Asia, offering a holistic approach of urban sustainability that is also reality-tested. From green buildings to smart city development, this book provides strategies on how different aspects and parameters of urban sustainability are linked and interact with each other. It is not only the procedural practice from the government and industry that are important for change, but also the focus on bringing people together to find common solutions that can effectively drive urban transformation. The authors spell out the importance for governments around Asia to provide leadership for sustainable growth, as well as the need for design solutions from individual buildings to district level integration to a smart city as a system. Lastly, the people-focused approach envisions sustainable communities and individual comfort that can be achieved by behavioural change, driven by trends toward low-carbon living.

It is therefore my pleasure to recommend this book to policymakers, built-environment industry practitioners, community-building professionals, and more generally, those who want to see green and liveable cities. It is my greatest wish that the sustainability practices outlined in this book can accelerate mindset changes and enable the green culture to become second nature, so that cities in Asia become more desirable places in which to live and work.

Andrew Chan Ka-ching
President, Hong Kong Academy of Engineering Sciences
Past President, The Hong Kong Institution of Engineers
Founding Chairman, Hong Kong Green Building Council
Chairman, Trustees Board, Arup Group

Preface

With the historical agreement of 195 countries at the twenty-first annual Conference of the Parties (COP21) in Paris in December 2015, climate change has been acknowledged as an urgent matter for everyone around the world. Climate change has come to the forefront of the global agenda, and is a challenge that has to be combated collectively by all countries. Since the ratification of the Kyoto Protocol in 1997, developed countries have put the agenda on the national level and are trying to formulate supporting policies to implement their respective commitments. Some positive action has already been taken – endorsing emission limits and conducting studies on the associated economic and social implications in an attempt to deliver the anticipated environmental benefits. The building and infrastructure market has been informed about the potential benefits of green businesses. Driven by the global sustainable development movement, the construction industry has responded with developing all kinds of new technologies aimed at reducing the direct emissions produced from power plants to vehicles, and at reducing indirect emissions by enhancing energy efficiency and green building products. The green movement has also spread to Asia and has been fuelled by regional economic integration. In a short space of time, the market created the hope that the target of creating a sustainable future could be realised.

However, two decades have passed and little tangible progress has been observed concerning the curbing of carbon emissions, let alone reducing the absolute emission level to below the 1995 level as committed to by many of the nations that ratified the Kyoto Protocol. Some countries have even gone so far as to pull out of the Protocol entirely. With the disappointment of COP15 in Copenhagen in 2009, the green movement combating climate change has lost some of its momentum. The talks on vulnerability and resilience to extreme climates and disasters are muted, and people and the market are reverting back to old habits and business as usual, wrongly assuming that the problem is over. This roller-coaster ride through these decades is hoped to be changed as the agreements signed at COP21 showed the commitments for both financial and carbon reduction targets from the countries.

There are many reasons for the loss of focus addressing climate change. Although many actions and initiatives have been undertaken worldwide, many of them are only on the policy level. Little effort has been made targeting and engaging people or transforming the industry itself. One fundamental question has not been addressed – how can the ordinary person, the individual, help? Without support from the public, practitioners, and the market, no policy action can be sustained. The gap between vision and reality is due to the misalignment of the priority of policies and actions. To engage the people, the right policy that addresses basic needs is required. The process of urbanisation has changed our lives and living environment more profoundly than at any other period in human history. In particular, the pace of urbanisation across Asia in recent years is unprecedented.

Nowadays, most people who live in high emission countries (the US, the EU, Japan, etc., except China which is still in the process of becoming one) live in cities. A city provides all the modern necessities such as education, jobs, public health, security, and so on. To sustain the operation of a city, a great deal of natural resources need to be consumed and in many cases wasted, such as fossil fuels for energy, forests cut down and replaced by farmland, and water for drinking as well as industry. Cities have created many environmental impacts and increased the load on the natural habitat. Air pollution, water pollution, and soil contamination from waste by the developments are amongst the most common. The living environments of cities have also changed significantly. In cities like Hong Kong, people live in a congested environment and buildings have gotten taller to cater to the density requirement. The basic elements for sustaining a healthy life such as good ventilation and adequate daylight have been jeopardised; improving buildings and the built environment is the crux of the solution for city dwellers and the environment.

To rectify the problem, we need sustainable development, and that requires the concerted efforts of all stakeholders. This is not a simple task, as the sustainability of buildings is a multi-faceted problem. We need collaboration and strong leadership to reboot the green movement at the national level, in the government and the private sector alike. But most importantly, we need the continuous support and action from the stakeholders on developing the right policies, as well as more innovation concerning our practices, and fundamental behavioural changes to our consumption for the betterment of the environment. Specifically:

1 **On policy** – we need to set priorities for our policies on sustainable development; we also need to lay out the roadmap and the action plan for long-term strategies; and the institutional arrangements to enable implementation of policies and strategies;
2 **On design** – the practices of our building industry need to be redefined, in particular applying new design strategies for green transformation;
3 **On people** – the mindset of ordinary people needs to be changed and must allow for their behaviour to change to a more sustainable living style, both at work and at home.

The following chapters have been written from the perspective of the people and building sectors of Asian countries to provide some answers that address these aforementioned issues; and to a lesser extent, some climate change challenges as well. Examples of Asian countries addressing the sustainability of buildings and the built environment were provided. They illustrate the valuable know-how of building design and the strategies of implementing policies. In particular, the Chinese examples (as a model) are of great importance globally, partly due to the scale of their carbon emissions and partly because of their impact on emerging economies.

It is envisaged that solutions for green transformation will ultimately come from changes to people's behaviour and the practices of our building industry in the green movement. The means must be a combination of appropriate polices aimed at facilitating the market forces working in conjunction with breakthroughs in technology that make green solutions viable and lasting.

Acknowledgement

First and foremost, we want to express our gratitude to Arup for the opportunity to dedicate time and effort to the topic of urban sustainability. Arup has provided an innovative and supportive environment and access to an extensive global network, so that more creative ideas can be generated and tested before the actual implementation. We are thankful to the company founder, Sir Ove Arup, and the management team in creating such a culture, which encourages us to make a broader social impact in what we do.

Next, we would like to thank our clients in trusting us to serve in their projects. We are glad that we have often had opportunities to go the extra distance in coming up with workable solutions that brought broader benefits to the projects and community. Project teams from different disciplines often worked together, and we are thankful for the friendships we have built.

In our own Building Sustainability Group in East Asia, we are glad to be able to partner with colleagues who share the same passion about sustainability. In particular, we would like to thank our team (Dr. Tony NT Lam, Dr. Camby Se, Jimmy Yam, Wai-Ho Leung, Dr. Kevin Wan, Mark Cameron, Derek Chan, Henry Au, Teri Tan, Eriko Tamura, and Tao Li) in providing valuable ideas for this book. Also, we thank our administrative staff (Mona Kwok, Melinda Chan, Thera So, and Raymond Chan) for their countless efforts and help.

At John Wiley & Sons, we want to thank our publisher, Dr. Paul Sayer, for seeing the value of this book and providing feedback on enhancing the content, and his colleagues, Viktoria Hartl-Vida and Monicka Simon, for following through the process to the book's completion.

Vincent Cheng would like to thank his wife and his two sons.

Jimmy Tong would like to express appreciation to Janet, Jocelyn, Josiah, and his parents, Shiu Hon and Mei Mei, for their love and support. In addition, Jimmy would like to thank Professor Ephraim Sparrow for his friendship and inspiration for professional excellence.

Hong Kong
June 2016

Section 1
On contexts

Chapter 1
Introduction

1.1 WHY SUSTAINABILITY MATTERS

The world has witnessed a rapid and unprecedented change in human activity over the past two centuries. The industrial revolution changed the way we live, work, and interact with the nature. The process of urbanisation as a result of this revolution has not only changed the global economic context but also created an environmental crisis, which was not recognised until recently. Climate change resulting from human activity is a known fact that is beyond question. The challenge confronting us now is how to resolve the problem.

Early discussions concerning the issues surrounding global warming have primarily centred on the perspective of the Western developed countries, in an attempt to clean up the mess that they created. Yet the world has failed to observe that answers should really be coming from the East, where more people live and where more countries are undergoing rapid urbanisation at a pace much faster than their Western counterparts underwent decades ago. The world should be treating the urbanisation process in Asia as an opportunity to develop a new model aimed at reversing the unsustainable processes of the past, and to devise solutions for future sustainable urbanisation.

In recent years, many Asian countries have proactively taken action to address the issues of climate change. These issues are multi-faceted and therefore the solutions are multi-dimensional. Sustainable development has been widely accepted as the solution to address the environmental pressure that results from rapid urbanisation. The building sector has been unanimously identified as the key area of focus in Asian countries, because more and more new buildings will need to be constructed in order to meet the growing housing demands and to support the economic activities of the increasing number of mega-cities that will be built over the coming years. There are huge disparities in the economic development of various countries in Asia. More developed countries have formulated strategies on sustainable development, whereas less developed countries are still at

Building Sustainability in East Asia: Policy, Design, and People, First Edition. Vincent S Cheng and Jimmy C Tong.
© 2017 John Wiley & Sons Ltd. Published 2017 by John Wiley & Sons Ltd.

a "soul searching" stage. These more developed countries have treated the issues of sustainability as an opportunity to enhance environmental performance and to address the demand for social equity. Externally, these countries have taken the green transformation as opportunity to enhance their competitive advantages on an international level as well as in the global market. Japan, Korea, and Singapore are perfect examples in this regard.

To sustain the green movement, it is important to put ideas into practice. Experiences in Asian countries have demonstrated the importance of multi-dimensional solutions. Effective policies that activate market forces are being formulated. Encouraging stakeholders to develop technical solutions is also critical to policy implementation. Strategically, Asian countries are undergoing a process that builds the capacity of their green power and takes advantage of economic improvement to lead sustainable development on the global front.

1.2   WHY ASIA MATTERS

With a population of 60% of the world's total, Asia plays a pivotal role in global sustainable development. Though many Asian countries are still at relative low level of development,[1] their rapid urbanisation in recent years has created a great deal of uncertainty regarding whether or not the world as whole can be successful in tackling the climate change issue.

The recent economic development of Asia is one of the greatest success stories in human history, with hundreds of millions of people working their way out of poverty. In particular, the East Asia economies of Japan, Korea, Taiwan, Hong Kong, with the exception of Singapore, and of course China have grown at a faster rate and for a longer period of time than the world has ever seen.[2] This is exemplified by the four-dragon miracle in the 1980s. In every aspect, many Asian cities have become well-developed, yet their economic success has come at a great environmental price. The development model in Asia is mainly resource-intensive in order to drive the growth of the economy as quantified by the gross domestic product (GDP). The main points of focus are on economic development and infrastructure investment so as to facilitate further development. China followed suit in the late 1990s but on a much larger scale, and in the process caused a great many environmental disasters in the region. According to the U.S. Embassy, Beijing's air quality is really bad. As of 2015, when this book was written, choking pollution regularly smothers the capital, and since 2010, China has been the world's largest carbon-emitting nation.

To satisfy the rising consumption of products due to the success of economic development in Asia, there is a growing demand for the resources used in sustaining industrial activity. In addition, to support the process of urbanisation, construction activity in Asia is the highest in the world in terms of volume and speed. This is in order to meet the never-ending demand for space for housing and work.[3] According to information from the United Nations Environment Programme (UNEP), China alone accounted for more than half the total of global construction activity in recent years. It is envisaged that this trend will continue to grow with more and more countries in the region following in the footsteps of China in their process of urbanisation.

The economic, social, and environmental sustainability of Asia is crucial if the world wants to see a continuous improvement in the quality of life for billions of people and for those city-dwellers whose hopes of a better future are not endangered by pollution or a poor infrastructure. The ultimate question is how best to provide room for the Asian urban population to live, work, and play as Asia gets richer and more populous, whereas at the same time energy, clean air and water, and living space are becoming ever scarcer.

Resource depletion and environmental factors are not new problems to many countries in the course of their industrialisation process. However, the discussion surrounding sustainability in Asia was intensified with the signing of Kyoto Protocol in 1992. More developed regional economies such as Japan have since begun to formulate policies to tackle the issue of climate change by setting emission targets in a similar manner to that of Western countries. Thereafter, policies have looked into the broader perspective of sustainability instead of simply focusing on energy security or tackling immediate environmental problems such as air and water pollution. Since 1992, initiatives dealing with fostering sustainable development have propagated throughout the region and other East Asia economies, such as Hong Kong, Singapore, Korea, and most recently, China have since followed suit with this green movement. For many of them, the sustainable development model provides an alternative for improving not only social assets but also economic development. The boom of green technologies enhances competitive advantages in the advanced technology market. Korea has been advocating this "low-carbon green transformation" initiative with a view to taking the lead in the global arena.

For those urbanising countries, new infrastructure and more new buildings are being built as more and more people are migrating to the cities. China, as the leader of these emerging economies, has taking advantage of the booming properties market to promote "green" buildings. Various control mechanisms were introduced over the past decades to regulate the design of buildings with regard to their energy performance. A new market of green building has also been established, where market forces capitalise on green designs. The building industry has begun practicing sustainable development by incentivising more green building technologies. This growth of the photovoltaic (PV) industry is a result of such favourable market conditions.

With more demands for green products (buildings and non-buildings alike) and for effective policies to facilitate market transformation, Asia has the potential to emerge as the leader of low carbon development in the coming years. Regional economic integration provides Asia with the opportunity to work in collaboration on reducing the barriers for green practices of production and trading, and enable agreement on cross-border carbon tax and carbon trading.

1.3   WHY BUILDINGS MATTER

Years of global discussion since the release on the Brandtlant Report in 1987 have come to the conclusion that implementing the principles of sustainable development is the only viable solution for addressing climate change. Rapid urbanisation has fundamentally changed the context of our living environment. High-rises and the compact city urban design approach are prevalent in Asian urbanism. Addressing the urban challenges of our built environment and exploring the opportunities for changes to how we build and use our buildings is imperative for green urbanism.[4]

1.3.1   Root causes and solutions to the problem

The building sector is an important component of sustainable development because it consumes more resources, in particular, energy, than any other sector. The building sector in the US consumed 75% of all electricity and 40% of all energy products.[5] It also consumed 40% of the raw materials and generated 30% of the waste for 2009. In the same year, the U.S. building sector generated 46.7% of all greenhouse gas emissions, far more than transportation or industry produced.[6] As many Asian countries are beginning to become urbanised, buildings will need to be designed and operated in a similar manner to that of the US if no new green standards are put in place. If not, the same types of problems can be expected in Asia in the near future.

Because of its massive size, the building sector offers significant opportunities in the reduction of greenhouse gas emissions and in providing the required solutions. Many studies worldwide have demonstrated that so far, green buildings are the low-cost option in obtaining a significant reduction in carbon emissions, that is, they are the "low-hanging fruit". For example, energy efficiency measures aimed at reducing building energy demand would be much more effective than the installation of increased power plant capacity designed to provide the "saved demand". Sustainable development is particularly imperative for Asia as more and more buildings are going to be built. In other words, fewer power plants using fossil fuels will be needed by the conventional practice of building sector. For example in China, with about 40% of its 2030 building stock yet to be built, implementing building codes for energy efficiency will yield substantial results. China is among the first non-OECD (Organization for Economic Cooperation and Development) country to introduce a mandatory energy code.[7] Most developing Asian countries now have their own energy codes to safeguard the energy efficiency of building design, a practice that has been implemented for years in Japan and other developed Asian countries.

1.3.2   Eco-city principles

The impact of buildings on sustainable development goes beyond the construction and operation of the buildings themselves. It also involves the use of land, the planning of infrastructures, and the provision of a quality urban living environment for people, for example, improved liveability. Eco-cities or green urbanism, with a focus more on environmental sustainability, is currently being promoted across Asia. The region is in need more than ever of green urbanism as it has more densely populated cities than anywhere else in the world. Currently, there are more than 200 mega-cities that have a population of more than 1 million. In particular,

China's development has provided an example for many developing countries on what to prioritise and the possible solutions for a more sustainable future. Over the last 30 years, China has experienced unprecedented economic development, with an annual growth averaging over 10%. Accompanying this growth, modern buildings, transport, and public service infrastructures are fast being built across the country. Rapid urbanisation is accompanied by a significant pressure to provide jobs and economic opportunities, housing, public services, and an improved quality of life. Given that cities contribute to more than 70% of energy-related carbon emissions, addressing cities' emission levels is a crucial part of reducing the economy's carbon intensity of China by 40%-50% by 2020, compared to 2005, the baseline year for comparison. The concept of eco-cities is becoming synonymous with the sustainable cities of the future.[8] China has begun to apply low-carbon city development concepts, becoming part of a global trend where different cities take a leading role in incorporating ecological and low-carbon development considerations into account with their urban planning and management models. The Ministry of Housing and Urban-Rural Development (MOHURD), has attempted to guide cities towards greater sustainability, including developing various eco-city standards and policies. A model for a high-rise and high-density eco-city is being formed. Moreover, eco-cities are currently evolving into smart cities that adopt digital technologies and Internet of Things (IoT) to make cities themselves more energy efficient. For example, the collection and analysis of large amounts of data (so called "big data") can help cities better understand building energy use and traffic patterns, and therefore distribute energy more efficiently.

1.3.3   Liveable spaces

The priority of issues concerning sustainable buildings varies amongst Asian cities because of the various disparities in their stages of development. In less developed countries such as China and India, resource efficiency is the top priority, followed by the basic provisions for addressing environmental problems, whereas in more developed countries such as Japan, Korea, and Singapore, liveability defined by a more pleasant, healthier, and comfortable environment for work and living is the core concern, higher in priority than resource use and the movement of people.[9] In Singapore, the provision of a liveable home is one of the principal objectives of its sustainability blueprint.

There is no universally accepted definition of liveability. A liveable space has to fit into the local social and cultural context and the expectation of the people. The physical environment as defined by the arrangement of buildings and public spaces can also improve the perceptions of people and allow them to be content with living conditions.[10] Specifically, high-rise and high-density compact city environments affect the liveability of many Asian cities. Improving the living environment is important to ease the overcrowded feeling of urban living. Quality, comfortable, and secure open spaces are also helpful in improving the sense of belonging in the community. The emergence of smart buildings with sensors designed to monitor and manage energy usage in public spaces and in homes can help save energy.

When it comes to the design of indoor spaces, health and well-being are the primary concerns across all cities in Asia. People who live in a city spend over 70% of their time indoors, either at home or at work. Healthy and comfortable environments improve the well-being of the occupants, and therefore, the productivity of employees. Recent pandemics in the region, such as Severe Acute Respiratory Syndrome (SARS) in 2003 raised government alert levels of the importance of hygiene, in particular at the community level. These are becoming more and more important criteria for corporations in the region when selecting premises for their offices or homes. Most of the green building standards have now incorporated these criteria in the assessment of building performance. New standards are incorporated into urban design to provide a healthy living environment. Air ventilation and micro-climate design are prevalent in the densely packed urban context of Hong Kong and Singapore. It is important that practitioners of the building sector can standardise urban design and take advantage of the benefits of good micro-climate design.

1.4   DIMENSIONS OF SUSTAINABLE DEVELOPMENT

Implementing sustainable development in cities is a complicated process and requires changes in practice in many areas. To address the multi-faceted issues of social, cultural, economic, and environmental factors, an over-arching policy and regulatory framework is required. Strong leadership from the government is a prerequisite, and this also requires support from other stakeholders such as business leaders, industrial practitioners, and the public. Collaboration, at both the local and international level, and partnerships amongst the stakeholders are the keys to success. To sustain such a green movement, the continuous engagement of the public and incentivising the operation of the market is also important. Finally, green business requires the standardisation of market practices by building the capacity of the key players.

1.4.1   Policy support

Governments play an important role in formulating the appropriate policies that will support the implementation of sustainable development initiatives across Asia. With a relatively short history and market of green practices being in their formative stages, it is particularly pivotal for success to have the correct policy instruments. In most East Asia countries, governments have their own high-level national plans concerning the objectives and issues of sustainability. Some countries like China and Korea have set five-year plan targets. These policies help to direct the market in green business or to change the behaviour of industry or individuals in reducing their environmental impact.

"Command and control" are the widely used policy instruments when implementing the policies. Law, regulations, and standards are all in place for consistent enforcement within the industry. Experience in East Asian countries has shown that the regulatory approach is effective only in the early stages. It is very expensive to execute orders and evaluate performance. Some countries have little resources with which to update their codes or standards. How to work in collaboration with industry has become the priority of certain governments when implementing green measures.

1.4.2   Green market and consumption

The idea of green building has been around for decades. In the early 1980s, one of the pioneers, Amory Lovins, an American physicist and environmental scientist, built a demonstration house at a 7,100-foot elevation in the Colorado Rockies. However, it has not been until recent years when sustainable development has become a key agenda worldwide that houses like his have appeared in the main-stream market. Green buildings only gained popularity in the property market when non-profit making organisations, such as the U.S. Green Building Council (USGBC), began promoting green certification which helped to spread the message of the benefits of green building and building green.[11] Leadership in Energy and Environmental Design (LEED), administered by USGBC, is by far the most successful certification and labelling tool in the world operated purely by market forces. It has created a demand for green buildings in the market and encouraged developers to build more green buildings. It has also provided the market standard for the design and construction of green buildings.

The development of green buildings in certain Asian countries has a long history. The first green building assessment code, Hong Kong Building Environmental Assessment Method (HK-BEAM) was implemented in 1996 in Hong Kong, well before the commencement of LEED in 1998. Other green building assessment schemes such as such as Comprehensive Assessment System for Built Environment Efficiency (CASBEE) in Japan were implemented in 2005, the Green Mark in Singapore in 2005, and the Ecology/Energy Saving/Waste Reduction/Health (EEWH) in Taiwan in 1999. The areas of focus for most of these assessment schemes are similar. Unique to context of Asia, some schemes such as Green Mark and BEAM Plus (an updated version of HK-BEAM) are not just market tools. They are supported by the governments through various forms of incentives, such as gross floor area (GFA) concessions or direct financial subsidies. The Green Mark is becoming mandatory for all new construction projects, and the Singapore government has set a target of having 80% of all the buildings to be certified by 2030.

As for green consumption, widespread adoption of sustainable construction requires momentum from businesses for more products and a standardisation of practices. Initiating sustainability because "it is right thing to do" is a nice idea, but typically it will not last, nor can regulations be sustainable. We need to build more business cases. In Asia, a number of businesses that provide green products are taking innovative measures in response to the environmental challenges they face.[12,13] Private firms are using technology, money, and above all, employee inge-nuity to begin solving Asia's environmental challenges. These activities are not "greenwashing", philanthropy, or "corporate social responsibility", but hard-headed business responses to opportunities born out of crisis. It is happening because there are business opportunities. Companies in Japan, Korea, Taiwan, and Singapore have all used their technical and engineering expertise to take advantage of Asia's need to support more people at a higher standard of living while using fewer resources. Studies have been conducted recently in Hong Kong and Singapore to identify the reasons why businesses should start building green.[14] It was found that "lower operational costs" and "higher building values" were the key reasons for building green, whereas the "perceived higher upfront cost" was an undeniable stumbling block that deterred building designers.

In some countries such as China and Korea, there is policy support from the government to enhance green business through research and development (R&D) and international collaboration. These activities were intended to increase market share and leadership in the cutting edge green technology market, both regionally and internationally.

1.4.3   Technology push

Technology has helped the green transformation of our building industry. More high-performance buildings were built in the past few years than ever before in response to the global green movement. The concept of Zero Carbon Building (ZCB) or Positive Energy Building (PEB) is now prevalent in Asia. Many low carbon or zero carbon buildings were built as demonstrations, first in Japan in 2008, then in Korea, Singapore, and Hong Kong. ZCB has proved to be technically feasible in certain types of buildings. Efforts have been made by many leading corporations in the region to commercialise their adoption in the wider market. In parallel, industrial practitioners are also working on buildings that are adaptive to climate change. Buildings are now more resilient to extreme weather conditions, such as hurricanes, floods, fire, and earthquakes.

Technology also helps to change the behaviour of individuals and corporations in regard to wasteful consumption of limited resources, in particular energy and water. In fact, the root causes of many environmental problems are due to irresponsible human behaviour. The current consumption habit of urban living is not a sustainable one. It requires more resources than planet earth has available to support the current lifestyle of Western developed countries. The developing Asian countries provide an opportunity to devise a model for a new sustainable lifestyle of green consumption for us to follow, so that the future generations might still be able to enjoy a quality living environment.

1.5   SUSTAINABILITY IN PRACTICE

The green transformation of the building sector requires the building industry to practice sustainability. Standardisation (i.e., transforming the current situation) of building practices requires the involvement of the professionals in the building industry as well as engagement with the public. This means firm collaboration and partnership between the private and public sectors on resolving issues such as regulations and standards, the design and construction of buildings, as well as public consumption behaviour.

1.5.1   History of green building in Asia

We only have a short history of two decades worldwide in the development of green buildings. It began in 1998 when most of the leading global green builders gathered at the first World Sustainable Building Conference in Vancouver. Since then, the movement of green building has gained a great deal of popularity. In Asia, the beginning of the green building movement was slightly later.[15] Prior to the 2000s, guidelines for sustainable buildings were scare in the tropics and sub-tropical Asian countries. The guidelines which primarily originated from the US and Europe did not necessarily apply to the region.[16] The initiative gained traction, however, when carbon reduction went to the top of the agenda in many Asian countries after the ratification of the Kyoto Protocol (Table 1.1).[17]

1.5.2   Capacity building – green professionals

The practice of sustainability as a discipline at the professional level is still in its infancy. It is a multidisciplinary practice requiring the integration of expertise from a variety of different professionals. Although many professional institutions for architects, engineers, and surveyors have a division of sustainability aimed at promoting the design of sustainable buildings, there is no one single body that oversees consistency of practices. Some global organisations, such as the World Green Building Council which was formed in 1999, were created to coordinate the development of green buildings and to enhance the collaboration among its members in different parts of world for the advancement of green building design. However, different countries have their own barriers that must be overcome and various institutional arrangements need to be established. More initiatives on the local level are required to develop the practice of green building.

In Singapore, the practice of green building or Environmental Sustainable Design (ESD) in the industry is booming. This is largely due to the implementation of Green Mark certification tools, which have been boosted by market forces. In China, it has recently become a mandatory requirement on sustainability for public buildings to facilitate capacity building. This has generated a professional practicing of *Building Sustainability*, a practice that demands integrated design and collaborative work from different professionals. Green professionals (Green Mark Managers in Singapore and BEAM Professionals in Hong Kong) also play a key role in the construction of green buildings.

Table 1.1 Key Milestones of Green Building Development in Asia

Year	China	Hong Kong	Japan	Singapore[18]	Taiwan
1996		Launch of HK-BEAM			
1997			Hosted UNFCCC (COP3) in Kyoto – Kyoto Protocol		
1998		Launch of HK Energy Efficiency Registration Scheme for Building			
1999			Second Revision of Energy Saving Standard for Buildings Launch of Top Runner Target Product Standards		Launch of EEWH
2000					
2001					
2002					
2003			Launch of CASBEE		
2004					Launch of Green Building Materials Labelling System
2005	Mandatory implementation of Building Energy Code by MOHURD		Organised SB05 international Conference	Launch of BCA Green Mark Certification programme	Green Building Basic Chapter of the Building Technical Regulations was announced
2006	Long-term Scientific and Technological Development Plan (2006~2020)	Launch of Comprehensive Environmental Performance Assessment Scheme for Buildings (CEPAS)		Launch of First Green Building Masterplan	At least 5% utilisation of green building materials for public buildings was required according to the Building Technical Regulations

Year					
2007	100 green building demonstration projects and was set in Eleventh Five Years Plan			Launch of Sustainable construction masterplan	The percentage of green building material utilisation was raised up to at least 30%
2008	10 projects obtained green building rating			Amendment of Building Control Act to impose minimum environmental standards	
2009	Launch of Green Building Labelling Scheme known as Three-Star, by MOHURD	Formation of Hong Kong Green Building Council; Energy Efficiency (Labelling of Products) Ordinance; HK Climate Change Strategy and Action Agenda	Launch of "Eco-point Programme" for Residential Buildings (ended in 2013)	Opening of BCA Zero Energy Building; Launch of Second Green Building Masterplan; Mandatory for all new public buildings to achieve Green Mark Platinum.	
2010			Launch of Tokyo Cap-and-Trade Programme	Launch of occupant-centric Green Mark programme	Intelligent Green Building Promotion Programme (2010–2015) was ratified
2011		Hang Seng Corporate Sustainability Index Series	Launch of LCCM (Life-cycle Carbon Minus) Residential Building Certification Programme		Opening of the Magic School of Green Technologies
2012		Buildings Energy Efficiency Ordinance come into full operation; Opening of CIC Zero Carbon Building; HK3030 Campaign Started	Law regarding Promotion of Low-carbonisation of Cities	1000th Green Mark buildings & 100th Green Mark Platinum building project milestone; Law requiring minimum sustainability standards for existing buildings, including the submission of energy consumption and building-related data, and regular audits and compliance on the cooling system efficiency	

(Continued)

Table 1.1 (Continued)

Year	China	Hong Kong	Japan	Singapore[18]	Taiwan
2013	Green Building Action Program launched by State Council of China	Mandatory implemented of Building Energy Code	Third Revision of Energy Saving Standard for Buildings; Launch of Building Energy-efficiency Labelling System (BELS)		Launch of LCBA (Low Carbon Building Alliance)
2014	The GFA achieved 170,000,000 meter square for Green Building Rating			Launch of the 3rd Green Building Masterplan and developed the inaugural Building Energy Benchmarking Report	Launch of BCF (Building Carbon Footprint Assessment System
2015	The design of Green Building head for the direction of internet-enabled	HK Energy Saving Plan; Hong Kong Climate Change Report 2015	CASBEE-City for worldwide use		
2016			The Act on the Improvement of Energy Consumption Performance of Buildings (Building Energy Efficiency Act); Incentive Measures (Voluntary) in effect	Opening of SkyLab – BCA's Rotatable Test Facility	

[18] BCA (2009). Singapore Leading the way for Green Building in the Tropics. Building and Construction Authority Singapore.

1.5.3   Sustainable change for the green movement

To sustain the green movement in reducing carbon emission requires more "buy-in" from the change agent – the people that can make the difference.[19] Sustainable behaviour requires educating and positively influencing the public. People are willing to make changes if the changes are easy or if they look like solutions that will lead to a better life. Lifestyles of Health and Sustainability (LOHAS) is becoming a marketable business model for property markets in the region. Some developers are more willing to engage with their clients and promote low carbon living. This is a positive sign that the traditional carbon-intensive building industry is reversing its direction and changing to more sustainable development. On the individual citizen's level, the carbon calculator has been developed as a tool to induce a change in consumption behaviour.

1.6   ORGANISATION OF THE BOOK

This book is about the recent progress in implementing sustainable development in particular some East Asia countries. Issues of sustainable development are multi-faceted and the solutions are multi-dimensional. Buildings and the building sector are the key aspects of this book. The authors will describe their own hands-on experiences through their project experiences in the region in order to elaborate on the principles and practices. This book is organised into four main sections, the context, the policy, the design, and the people.

Firstly, we will look into the context of Asian sustainable development and the problems that arise. Then in Chapter 2 we will discuss the rapid process of urbanisation. Chapter 3 explores the challenges, as a result of this process. Chapter 4 will look at how international initiatives are helping to drive the green movement forward and solve the problems.

In the section concerning policy, we will discuss the major policy objectives and priorities of these selected Asian countries in East Asia in handling the key challenges and how they turn the challenges into opportunities. Chapter 5 will discuss the policy agenda, whereas Chapter 6 will look into the implementation details of the policies.

In the section covering design, we will look into the technological solutions available to these East Asia countries. Chapter 7 focuses on the design issues of sustainable development. Chapter 8 is a detail account on the engineering solutions of the key design issues. Chapter 9 will then be about the various de-carbonisation strategies.

In the section on people, we will discuss how the issues of sustainability affecting people and how they can help to make the necessary changes. Chapter 10 will elaborate further on the unique high-rise and high-density environments that people are now living in. Chapter 11 will provide more examples on how Asian city dwellers are creating sustainable communities. Chapter 12 will provide cases studies on how people can play a key role in the solutions by being the agent of change for the entire green movement. Finally, this will be followed by the conclusion in Chapter 13.

Chapter 2
Rapid urbanisation

2.1 INTRODUCTION

It is a fact that nowadays climate change is primarily due to the human activity of the last few centuries. The process of urbanisation which resulted from the economic and social development of the time has changed the environment of many countries, developed and developing ones alike, with regard to their demographics, economic activities, and the consumption behaviour of their populations. In recent years, the world has witnessed dramatic changes in China because of such rapid urbanisation. On one hand, urbanisation has helped to lift millions of people out of poverty, but on the other hand it has created a great many negative impacts on the environment.

The process of urbanisation in Asia will continue for many more years to come, as other less developed countries also want to follow in the footsteps of China and enjoy the benefits brought about by economic development. With the size of the Asian population, the foreseeable results will be detrimental to the environment if no action is taken to implement sustainable development across the region. The problems facing the global and local environments will become even more challenging for future generations. It is imperative for us to understand the trends in the urbanisation of Asian countries and identify possible solutions to change the course before it is too late.[20]

2.2 ASIAN URBANISATION IN CONTEXT

Cities are instrumental to the economic growth of a country, and are pivotal in the development of other components of modern society, such as culture, education, healthcare, and political power. In fact, all developed countries are highly urbanised.[21] Urbanisation in Asia is unprecedented because of the scale of the changes that have occurred – the number of countries, the number of cities, and the number of

Building Sustainability in East Asia: Policy, Design, and People, First Edition. Vincent S Cheng and Jimmy C Tong.
© 2017 John Wiley & Sons Ltd. Published 2017 by John Wiley & Sons Ltd.

people involved. It is not difficult to understand how urbanisation has brought about enormous economic and social changes to various Asian countries and how it has benefitted most of the people living in the region. Regarding social development, while more developed cities can provide sufficient housing, social services (healthcare and education), maintain a liveable environment, and develop effective systems of governance and management; on the other hand social inequity and inequality are not uncommon in some of the less-developed countries. Economically, cities create the platform for the trade and services industry and enhance economic development. Rapid growth in gross domestic product (GDP) and an increase in disposable income are amongst some of the direct benefits of urban economic development. As a result, more and more people are attracted to cities because of better working and living conditions, and the cities themselves are growing bigger and bigger.

According to the United Nations Economic and Social Commission for Asia and the Pacific (ESCAP),[22] urbanisation trends in Asia and the Pacific have some distinct characteristics. Firstly, its urban population grew faster than any other region in world. Secondly, it has more than half of the world's mega-cities to accommodate the increasing population. Thirdly, economic growth is primarily led by cities in the region. Fourthly, urban growth is not environmentally sustainable. In response to these drivers of change, a model for green urbanism is being formed in the Asian cities where resource efficiency, resilience, and sustainable consumption are the key aspects of growth.

2.3   DEMOGRAPHIC CHANGES

Urbanisation triggers changes in demographics, and this is characterised by population size, urban population growth, and the shifting of its age composition. These changes influence all aspects of social and economic development in cities, including providing job, healthcare and education, and are connected to the use of land, water, air, energy, and other vital resources. It is therefore important to identify the interaction between demographic processes, natural resources, and life supporting systems, and start planning for various demographic changes.[23] The objectives are to capitalise on those demographic dividends to boost growth, build the community, and minimise the environmental risks.

2.3.1   Global population trends

Since the end of World War II the global population has expanded rapidly. In the 1950s, the global population was 2.5 billion. By 2013 it has almost tripled in number to 7.2 billion. By the year 2050, it is projected to be four times that number at 9.6 billion. The global population is expected to grow at a rate of 1% per year, gradually dropping to an annual increase of 0.5% by 2050. In terms of the numbers, the increase will be approximately 80 million per year from 2010-2020, and then 40 million annually until 2050. Amongst the top 10 countries with the highest populations in 1950s, Asia accounted for 4 of them: China, India, Japan, and Indonesia; in 2013, it increased to 6 countries: China, India, Indonesia, Pakistan, Bangladesh, and Japan. It is forecast that by 2050 Asia will still account for 6 most populated countries: China, India, Indonesia, Pakistan, Bangladesh, and the Philippines) (Figure 2.1).

Asia has become the major percentage of the world's population. In 1950 it was 55%, equivalent to 1.4 billion, in 2010 it accounted for 60%, and it is projected to be 54% by 2050, equivalent to 5.2 billion or 3.77 times that of what it was in 1950 (Figure 2.2). The rapid population growth in Asian countries has inevitably driven to the need for economic development. The change in mode of the economies of Asian countries, that is, the processes of industrialisation, commercialisation, and modernisation, has brought about a substantial need for urban development within and outside the region through regional integration and globalisation respectively.

The size of the population in Asia and its growth will result in a strong impact on the outlook of global sustainability. On the one hand, the population can provide the labour force for the process of industrialisation in the region and can help to address the issue of poverty that has plagued the region for a long time. On the other hand, urbanisation (if it still adopts a resource-exploitative approach) will create many environmental problems and worsen global resource depletion if proper practices of sustainable development are not in place. Because of the huge size of the population in Asia, the world is facing more uncertainty regarding sustainability as a result of rapid urbanisation.

2.3.2   Urban population growth

The size of the urban population stems from a natural population increase, urbanisation, rural-urban migration, immigration and so on.[24] Today, the number of urban residents is growing by nearly 60 million every year. Figure 2.3 demonstrates that the percentage of the population living in urban areas has continuously increased. In 1950, the global urban population was 0.8 billion, that is, 30% of the global population at that time. In 2015, it was 3.9 billion, that is, 53% of global population. The global urban population is projected to be 6.7 billion in 2050, that is, 70% of global population will be living in urban areas. It will be almost double the urban population of 3.4 billion in 2009. Population growth is positively related to the percentage of the urban population as more and more people move from rural areas to the cities for livelihood reasons and for a better quality of life. Greater infrastructure, more housing and social support facilities such as schools and hospitals need to be in place in order to support urban living. Without proper planning and governance in place, uncontrolled urbanisation might create social instability and further environmental problems.

In Asia, the percentage of the urban population was around 16% in 1950, it has now increased to 40%, and is projected to be 65% by 2050. This makes an increase rate of 1%-5% annually. Japan, Korea, Singapore, and Taiwan are the leading economies in the region, and there is a pattern to the change of their urban population. Their urban populations started to grow shortly after World

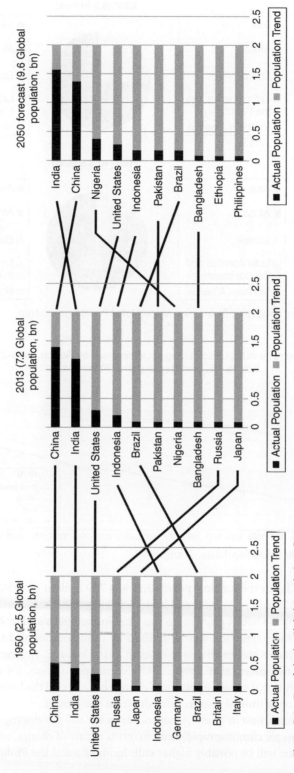

Figure 2.1 Trends in the Global Population. © Arup

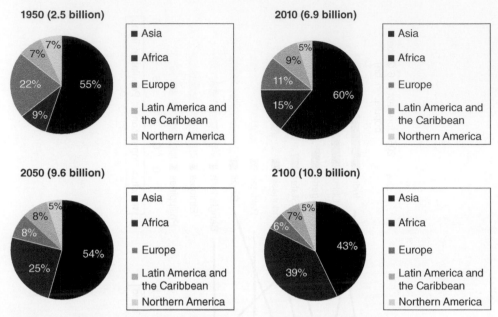

Figure 2.2   Trends in the Geographic Distribution of the Global Population. © Arup

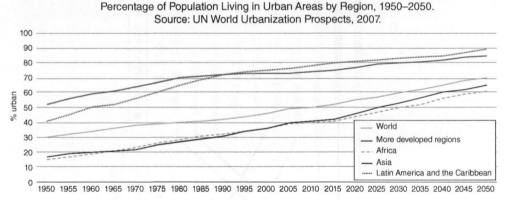

Figure 2.3   Trends in the Urban Population. © Arup

War II. In Japan the national population was 750 million in 1950; it boomed and peaked at 128 million in 2000, and is predicted to gently drop to around 100 million by 2050. Japan's urban population was 76% in the 1980s and in 2015 it was 93%. In Korea, the population was 25 million in 1960 and doubled to 50 million by 2013, of which 82% were living in cities in 2014 (Figure 2.4). Singapore and Hong Kong are two places that have a 100% urban population, 5.4 and 7.2 million respectively in 2014. Such a pattern of change has affected the labour structure and the planning of limited resources for different age groups.

Most South East or South Asian countries are still developing countries. Their populations are climbing rapidly and in terms of rate of change, and the process of urbanisation will be possibly higher still. Indonesia and the Philippines are about

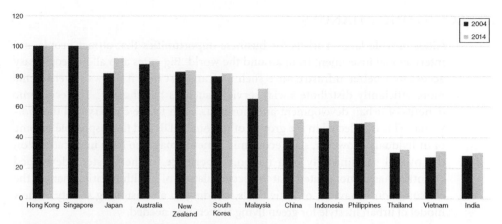

Figure 2.4   Urbanisation Rates of Asian Countries. © Arup

50% urbanised and they are projected to be 60% and 65% respectively by 2050. Whereas Thailand and Vietnam are at present about 30-40% urbanised and are projected to be 55% by 2050. The experiences of the regionally developed countries can help to shed light on effective planning for social development.

The carrying capacity of the earth to accommodate its population is a limited one. Urban consumerism has raised the demand for energy and other resources per capita, and therefore, there are implications for the three principals and essential stocks of renewable resources, that is, forests, grasslands and fisheries. The utilisation rate is faster than the replenishment rate and faster than the current population and living standards can cope with. Preferably, population growth should stop at some point, or else the world will become overcrowded. In fact, competition for freshwater sources has already created conflict between some countries in Asia.

More importantly, where will people live in the urban area while the population is ever increasing? Housing capacity will either expand horizontally, for example, a bigger city; or vertically, for example, a taller city. It creates huge pressure on social supports and environmental constraints.

2.3.3   The challenge of an ageing population

Urbanisation has also created social challenges due to a demographic change. Some Asian developed countries are facing the problem of an ageing population. Japan's ageing population outweighs all other nations in the world, as the country is purported to have the highest proportion of elderly citizens. According to 2014 estimates, 33.0% of the Japanese population is over the age of 60, 25.9% are aged 65 or above, and 12.5% are aged 75 or above.[25] An ageing population not only affects the productivity of a country but also creates new social issues such as adequate healthcare and a pension system for the aged. The onset of rapid ageing is an early warning signal and an urgent call for action in Asia.[26] Asian cities need to adapt to older citizens through better urban design that takes into account quality of life and a healthy environment for these citizens.

2.4   ECONOMIC CHANGES

Cities provide large markets for business opportunities that attract people and international investment from around the world. Big cities also allow people easy access to a better infrastructure such as transport and it allows governments more efficiently distribute social services such as healthcare.[27] The economic benefits of urban development are enormous, both to the country and the individual. This is reflected in the growth of national GDP and the disposable income of individuals. However, such economic benefits may also have implications on the overall sustainability of a nation. Experience in the Western developed countries has demonstrated that the accumulation of wealth in cities changes the consumption behaviour of individuals, alluding to a consuming society. A new model of urban lifestyle for green living is therefore needed.

2.4.1   Growth in GDP

It has been well established that the size of the GDP of a country is highly related to its level of urbanisation.[28] The rapid process of urbanisation in Asia over recent years has boosted its GDP considerably. According to Asian Development Bank, by 2050, Asia's share of global GDP will increase to 52%, regaining the dominant economic position it held before the industrial revolution some 300 years ago. By that point, Asia's per capita income could potentially rise six-fold in purchasing power parity (PPP) terms, reaching today's European levels. Some 3 billion additional Asians will become classed as affluent by current standards.

Indeed, urbanisation has brought about significant economic benefits to many countries, in particular for China over recent decades. In 1978, the GDP increased from US\$ 148.2 billon to US\$ 8.2 trillion by 2012.[29] Although slowing down to about a 7% growth in 2014, it was still the fastest growing economy in the world. If this development trend continues, it is estimated by the International Monetary Fund (IMF) that the GDP of China may reach US\$ 33.5 trillion overtaking that of the US, the largest economy in the world, by the year 2019 (Figure 2.5). Economic development provides the capital for transforming the built environment, and therefore living conditions and people's lifestyles.

2.4.2   Increased income

One of the positive outcomes of a rapid growth in the economy is the increase of people's income. East Asia is the fastest-growing region in recent years. From 1950, income increased sharply, first in Japan and later in Hong Kong, Singapore, and then Korea, Taiwan, and China.[30] Specifically, the GDP per capita of China increased three-fold during the process of rapid economic growth from below US\$ 1,000 in 1987 to US\$ 6,000 in 2013. Although still not considered as a "well-off" country, urbanisation has brought millions of people out of poverty. This increase has also fuelled the process of further rapid urbanisation. The results in the Asia Pacific region have shown a non-linear growth of income with urbanisation rates.

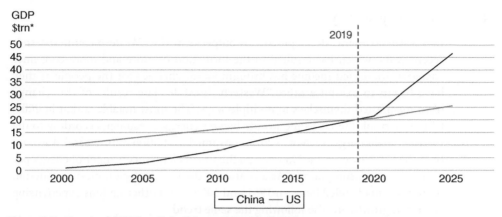

Figure 2.5   Trends of GDP Growth in China and the US. © Arup

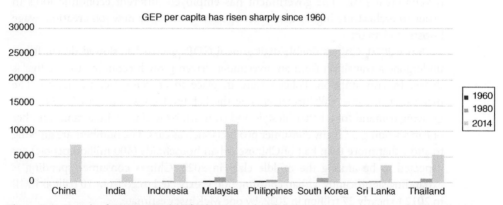

Figure 2.6   Trends of GDP per Capita of Asian Countries. © Arup

There is a threshold of urbanisation rates in counties before the increase in income is accelerated. This figure is at about 60%.

For some ASEAN countries, the GDP per capita has also risen sharply since 1960. Industrialisation began in ASEAN countries in the 1980s. The GDP per capita of South Korea and Malaysia was around US$ 2,000 in 1980 and rocketed to US$ 25,931 and US$ 11,387 respectively in 2014, representing a growth rate of 13 times and 5 times over the period. The GDP per capita of Thailand, Indonesia and Philippines were US$ 5,450, US$ 3,417 and US$ 2,935 in 2014 respectively, they are now 600%, 120% and 500% increased against what they were in 1980 (Figure 2.6).

The increase in income in cities is even more remarkable. In China, according to figures from the National Bureau of Statistics of China (NBS), urban disposable income climbed rapidly from 2,000 RMB per month in 1991 to 16,000 RMB in 2008, or in other words, it saw an eight-fold increase. It reached its peak of 24,565 RMB in 2012. It is expected to increase the disposable income of both urban and rural residents by over 7% annually, while the rural net income increased five-fold from 1,000 RMB to 5,000 RMB for the same period.

2.4.3   Consuming society

East Asian economic development has outperformed its Western counterparts in recent years, which has resulted in the formation of more and more consumer societies. Increased income has brought about changes for the people as they have begun to exhibit a more "Western" lifestyle. The results of this change have manifested themselves in multiple ways. The most notable is that in 2009, China overtook the US as the world's biggest energy consumer, according to the International Energy Agency.[31] China consumed 2.3 billion tonnes of oil, 4% more than the US. However, the US remains the world's largest consumer of energy per capita. Meat consumption in Asia is also on the rise. In Japan, meat production more than quadrupled between 1961 and 2005,[32] and other nations experiencing economic growth are also following the same trend.

China's National People's Congress deployed the twelfth Five-Year Plan (FYP, 2011-2015) and approved the new strategies of the national economic and social development plan. The government has employed different economic tools in order to orchestrate higher consumer spending and spur new job creation, wage boosts, and so on.

Since 2010, China's double-digit annual GDP growth has slowed down. It has undergone a transition from an investment-driven growth economy to one that is driven by consumption. Policies were in place to encourage consumption. The increase in income will further change the habits of consumption and the ever-growing demand for a better lifestyle and comfort. Nowadays, China ranks number one in spending on many consumer products such as cars. The numbers are growing to show that more than half of Chinese urban households (400 million people) are expected to be among the middle class in 2022. China's consumer spending is expanding significantly. Urban consumption is expected to go from 10 trillion RMB in 2012 to nearly 27 trillion in 2022, by one McKinsey estimate.

Increased production and consumption are driving energy use, CO_2 emissions, other forms of pollution, and the degradation of scare resources such as water and forests. A vicious cycle of urbanisation leading to environmental degradation is being formed (Example 2.1).

Example 2.1   Rising car ownership

One direct consequence of people's affluence is the ownership of private cars. In particular, the mode of urban development in Western countries has demonstrated that people living in cities tend to rely on vehicle transport for daily commuting if public transportation is not well developed. Today, there are around 1.2 billion passenger vehicles worldwide. By 2050, this figure is projected to reach 2.6 billion. The trajectory of vehicle ownership is quite similar to the one in the developed countries. Although relatively lower in the rate, the total amount is much more significant due to the much larger population. This has helped to build a huge car industry in China by 2013. Motor vehicle sales in China in 2014 were 23 million, 10 times that of 2001. The actual increase has been overwhelming. Annual passenger car sales have already overtaken the US at a sales rate of over 20 million cars per year. In 2018, China's vehicle fleet will overtake that of US, and will become the top country for total number of cars.

However, road and highway planning in China has not been able to meet the demands of road transportation. For instance, in Shanghai between 1995 and 2004, the average time it took to commute to work increased by 70%. Congestion on the roads causes billions of dollars of economic losses annually. Alternative transportation options for citizens must be developed and encouraged and a more sustainable approach to urban design must be put in place.

From Figure E2.1.1, besides the Chinese car market, the total vehicle and motorisation index of India and other major ASEAN countries will also climb dramatically. The scenario for India is equally impressive. Total vehicle ownership was 50 million in 2005, 120 million in 2005 and is projected to be 380 million in 2035. This is growth of 7 times over the period. India's motorisation index (the number of vehicles per thousand people) was 50 in 2005, 100 in 2015 and 280 in 2035, a growth of 3.5 times over the period. Major ASEAN countries will see stable growth. Total vehicles owned have increased from 75 million in 2005 to 140 million in 2015, and the number is projected to be 230 million by 2035, this is a three-fold growth, whilst the ASEAN motorisation index was 180 in 2005, 280 in 2015 and is projected to be 370 in 2035, about a 2 times increase over the period.

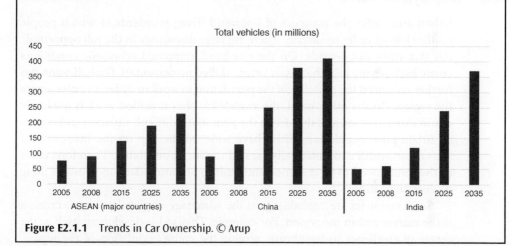

Figure E2.1.1   Trends in Car Ownership. © Arup

2.5   SOCIAL CHANGES

Urbanisation is more than just a technological or economic process. It is also a social transformation to a more modern way of life. It is a transformation of culture, behaviour, social institutions, and social structures. It affects how we work and live in a more organised and interactive societal system. Adjustments for rural individuals and families to an urban lifestyle are immense. Problems in access to housing and the labour market are particularly challenging for Asian cities.[33]

2.5.1   Housing needs

Housing is an important sustainability issue for Asian cities. Except for the developed East Asia economies, urban populations in Asia live in sub-standard housing. Most dwelling units are overcrowded with little more than 10 m^2 of floor space per person. In less-developed countries, sanitation, access, and a lack of reliable water and electricity are acute problems. Many countries have public housing in place, yet with the exception of Singapore, all have failed to provide adequate low-income housing, and private housing is not affordable to most people. This undermines social stability and is becoming a political issue.

The demand for housing in Asia will continue to grow with the rise of incomes and population growth. By 2030, an additional 400 million dwellings will be needed in Asia. Huge investment in buildings, infrastructure, and social services facilities such as schools and hospital are required. This poses a significant burden financially as well as on the environment, because green building practices are generally not in place in most of these developing countries.

2.5.2   Employment needs

Urban areas offer the prospect of improved living standards to which people require jobs in order to sustain. There are large disparities in the job opportunities that cities can provide. On the one hand, improved economic conditions create high-paying jobs that attract more skilled professionals. Globalisation has further enhanced the mobility of the population to work in different cities across the region. On the other hand, the availability of non-skilled work is gradually diminishing in cities, and this drives the poor into a difficult situation and creates an urban poor population. Cities have widened the gap on the inequality of income or wealth and potentially created the social instability. Therefore, cities of the future need to be designed and built to accommodate and sustain these diverse populations.

Some countries are capitalising on the advantage of globalisation to create a niche market within the region. For example, Thailand is developing itself as the centre of excellence in healthcare, serving the entire population of the region. Similarly, India is becoming the information and communications technology (ICT) giant in the region. More effort is needed for the emerging economies to create more jobs for their growing populations.

2.6   NEW GROWTH MODEL

Rapid urbanisation has exerted pressure on Asian countries in developing more cities for people to migrate to from rural areas. In China alone, by 2050 there will more than 210 cities with populations larger than 1 million. Cities will need to double their infrastructure when providing basic municipal services for water, electricity, sewage, and so on, to sustain its normal activities. The major challenges are proper housing and the necessary infrastructure for transportation, communication, water supply and sanitation, energy, commercial and industrial activities in order to be able to meet the needs of the growing world population. As a result a new model of urban development is required.

2.6.1   Mega and compact cities

Asia is the largest region for potential urban population growth and urbanisation. In 2014, there were 13 mega-cities;[34] this was equal to half the number of them in the world. The number of cities, both medium-sized (1-5 million) and large-sized (5-10 million) are densely located in the region.

China is the most populous country worldwide. Its population was 0.5 billion in 1950, 1.3 billion in 2013 (2.6 times that of 1950) and is projected to be 1.6 billion in 2050 (3.2 times that of 1950). The urban population in China has also been growing rapidly in terms of speed and scale. The percentage of urban areas was almost insignificant in 1978 when the country was beginning its process of modernisation. In 2010, the urban population raced ahead of the rural population. (By 2050, approximately 78% of the total population will be living in urban areas.) A statistics from the McKinsey Global Institute suggested that the urban population will increase from 63 million in 2010 to reach 990 million by 2030 where medium-sized and large-sized cities (larger than 1 million) are likely to increase from 153 to 226 over the period.

Other neighbouring countries in East Asia, such as Japan, Korea, Taiwan, and Singapore, also have a high rate of urban population growth. Cities have to expand and mega-cities must arise. By 2020, as shown in Figure 2.7, the population of Tokyo will be 37 million, and in Osaka-Kobe and Shenzhen that number will be 11 million.

Rapid urbanisation would drastically increase the demand for buildings and increased infrastructure which would lead to larger emissions during their construction and subsequent operation. In turn, this would exacerbate climate change. Environmentally sustainable buildings and infrastructure will be the key to urban resiliency.

The prevailing measures in Asia are to promote compact cities. These avoid urban sprawl and can make operation the most efficient as well as dramatically decreasing traffic congestion and greenhouse gas emissions. Countries such as China, India, Japan, and the Republic of Korea have all begun pursuing compact and smart cities, including reducing per capita carbon emissions.[35]

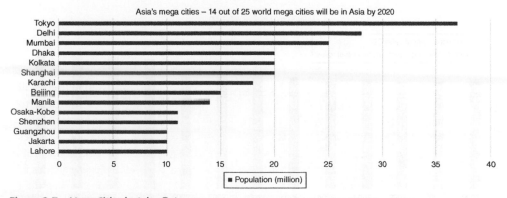

Figure 2.7   Mega-Cities in Asia. © Arup

2.6.2   Green building markets

In Asia, the rapid increase of income has also created a boom in the property market. People can now afford to own one or more properties. The living conditions of new flats are usually better and in some cities, the booming property market has helped to transform the urban environment in a much short period.

According to Oxford Economics, in Figures 2.8 and 2.9, construction market output increases in pace with population growth. China, India, Indonesia, and the US are the top construction markets in the period from 2005 to 2030. These countries are also ranked in the top 10 countries of highest population globally. This is because of the growing demand for living, social, and commercial activities in the cities and urban areas.

Building is a key sector in China's economy and continues to provide spectacular economic income. The value of construction output accounted for 26.1% of China's GDP in 2012, up from 18.8% in 2005, according to the NBS. Over the five years through 2013, revenue for the building construction industry has been growing at an average annualised rate of 22% to $1.54 trillion. There are about

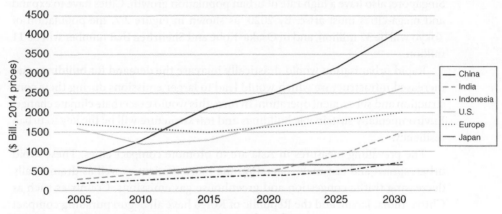

Figure 2.8   Market Sizes of the Top Construction Markets. © Arup

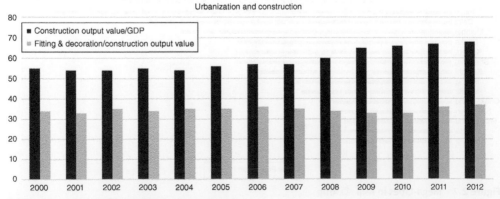

Figure 2.9   Construction Market in China. © Arup

27,000 construction firms in this industry, up from 22,000 in 2008. It is envisaged that the building market will remain as the most outstanding business sector and has a lot of room to create a sustainable built environment and to drive towards a low carbon community.

In China alone, the construction output value has been climbing across all types of buildings. The share of residential construction is most remarkable. Construction output value was 9,000,000 million in 2010 and is projected to be 25,000,000 million by 2019, this is equivalent to a 2.5 times growth rate. Residential and commercial construction values make up for more than 50% to 68% of the total construction value over the same period.

The large property market and the vibrant construction activity provide the opportunity to delivery green construction in China. With the right policies and government support, the country can attract the required capital to build more green buildings and infrastructure and build the capacity for sustainable development. Other countries in the region have also followed such green practices in the construction industry.

2.7   SUMMARY

Rapid urbanisation in Asia has created many economic miracles in the region. Asia is becoming the most important region for production of goods and for doing business. With more and more capital and talents moving to mega-cities in the region, their importance will be further strengthened in the years to come. Asia has the potential to be the leader in sustainable development as more new buildings and infrastructure are being built. Yet, many countries in region are facing social and economic changes that need collective effort to resolve. It needs the right governance in managing the cities and the right leadership to change the behaviour of businesses as well as the individual. It is important to not allow the region to become just a consuming society.

Chapter 3
Urban environmental challenges

3.1 INTRODUCTION

Urbanisation in the past few decades has accelerated human activities for industrial and commercial productions, which has exacerbated a number of environmental problems. In Asia, regional urban economies (Japan, Korea, Taiwan, Singapore, Hong Kong, and China to a lesser extent) have developed through environmentally exploitive models in their urbanisation process. As a result, many cities are now confronted with immense environmental challenges and some are facing multiple crises of liveability.[36] Access to clean land, water and air are in jeopardy for those cities that are still at an early urbanisation stage. Environmental crises from air pollution, water contamination and waste generation are prevailing and affecting daily life, in particular, the poorest of the poor and societies without social equity.

In parallel, cities are also facing adverse effects of climate change, which is escalating year by year and affecting everyone on earth. The risks of catastrophic natural disasters such as flooding, hurricanes or drought will severely impact regional developing countries that are less prepared. Buildings, infrastructure, government and social support facilities are vulnerable to extreme weather. The cost of repairing the aftermath of a natural disaster is prohibitively high for less affluent countries.

With more and more developing Asian countries undergoing urbanisation at a rapidly increasing rate, the outlook of resolving environmental and climate change problems are not promising. The world is facing more unprecedented challenges in its history. It is imperative that we understand the problems and their root causes, and collaboratively, take action to resolve them.[37]

Building Sustainability in East Asia: Policy, Design, and People, First Edition. Vincent S Cheng and Jimmy C Tong.
© 2017 John Wiley & Sons Ltd. Published 2017 by John Wiley & Sons Ltd.

3.2   URBAN CHALLENGES IN CONTEXT

People move from rural areas to cities in search of a better livelihood and an improved quality of life. A city not only provides the basic necessities of shelter, food and water but also offers working opportunities to meet the enjoyment of urban consumption. Social security and community support facilities can meet the needs of healthcare and educational requirements. Yet, these expectations do not come naturally, and a worse-off situation is not uncommon in many cases in Asia, where people often have to live in urban slums, suffering from the polluted environmental conditions of air, water and land. It has been argued that urban environmental challenges are the price of urbanisation as this has occurred in all developed countries during their urbanisation process. Some accept that these problems will only appear at the early stages of economic development and disappear when a country becomes wealthier. Yet, a "develop first, clean up later" strategy may not be workable for the current developing countries of the region due to two reasons. Firstly, the pace of urbanisation is unprecedented and the environmental consequences could be too difficult to clean up later on. Secondly, compounded to the local environmental challenges is the global climate change. The combined effects could be disastrous.

Globally, there is little agreement on defining urban environmental challenges as different regions/countries face different problems. Based on the scale and extent, urban challenges can be characterised in three types:

1 Global climate change
2 City-regional environmental challenge
3 City-local liveability deterioration

Climate change is a newly emerging challenge that is threatening all mankind. Asia and the Pacific in particular are regions that are most affected by natural disasters. Earthquakes, flooding, and tsunamis are known threats, and global climate change causes countries at a higher risk to be even more vulnerable. The impact of natural disasters on urban areas of regional countries can be devastating. Urban agglomerations with high population densities are experiencing high mortality levels in the case of natural disasters.

Problems of urban environmental degradation affecting the daily lives of city dwellers are equally daunting. Air and water quality is threatening millions of people in the urban setting, causing significant economic losses every year. Even the waste problem is becoming a key agenda for governments and policy makers.

In principle, enhancing liveability is in the interest and under the jurisdiction of a city. It has the authority and incentive to make the city more appealing to its citizens. The being said, many developed cities in East Asia region are facing degradation in built-up environments because of ineffective urban planning and poor management of urban facilities. There are too many concentrated and built-up areas causing heat build-up within cities and a lack of greenery to improve the quality of the ecological system.

These challenges have to be worked out from different levels and by targeting different stakeholders. Better urban practices and governance could help reduce the burden if not eliminate the problems (Table 3.1).

Table 3.1   Urban Challenges

Scale	Challenges	Typical issues	Causes
Global	Climate Change	CO_2 emission global warming; extreme weathers	Urbanisation; mass-consumption lifestyles
City/Regional	Environmental degradation	Air, water, and waste pollutions; energy and resource security	Rapid urbanisation
City/Local	Liveability deterioration	Urban Heat Island; Air ventilation; microclimate	Poor urban planning

3.3   CLIMATE CHANGE CHALLENGES

The process of urbanisation has created many challenges to our environment; many of which have not ever been encountered in the history of mankind, such as CO_2 emissions causing climate change. Over the past few decades there has been a broad global recognition that environmental resources are finite, fragile and exhaustible. Excessive environmental loading caused by man-made activities to sustain our urbanised living standards are now beyond our control. Climate change has resulted in extreme weather conditions and has caused many cities in the region to become vulnerable and their dwellers to suffer. The term "climate extreme" describes the immense impact of this challenge more so than the commonly used term "climate change".

3.3.1   Vulnerability to extreme weather

In cities, urban systems (including infrastructure, buildings and social support facilities) that are highly exposed, sensitive and less able to adapt are vulnerable to climate change.[38] Vulnerability is established as a function of exposure to climatic factors, sensitivity to change and the capacity to adapt to that change. The impact of climate change is profound and diverse. Melting glaciers, decreasing water volume in major rivers, rising sea levels, loss of biodiversity and an increase in natural disasters are posing a serious threat to people living in low-lying coastal areas. Developing countries will be hit more severely than developed countries because their infrastructure is not prepared for severe or extreme conditions such as storms or rising sea levels.

For coastal countries in Asia, serious damage, erosion and coastal flooding will occur at various locations following the passage of a typhoon. The situation will likely be aggravated by the long-term rise in sea levels and/or an increase in storm frequency due to climate change. Under such circumstances, the frequency of occurrence of extreme sea levels and coastal flooding may increase. A recent study investigated the potential implications of the effects of climate change on coastal structures in Hong Kong. The investigation ascertained the necessary revisions to the current design standards to minimise the risk of coastal flooding.[39] The study found that:

- Sea Level Rise: The combined effect of thermal expansion of oceans and a temperature rise causing ice sheets and glaciers to melt has resulted in a long-term rise in sea levels. The Intergovernmental Panel on Climate Change (IPCC) has modelled various scenarios in the mean rise of global sea levels, and the latest update was the IPCC's Fifth Assessment Report (AR5), which was issued in 2013. Based on this data, the analysis of Hong Kong's situation showed that the mean sea level of Victoria Harbour, on average, rose at a rate of 29 mm per decade since 1950. Based on the AR5 scenario Representative Concentration Pathway (RCP) 8.5 (W/m²), it is projected that the mean sea level will rise by about 14 cm and 26 cm at Year 2030 and 2050 relative to 1989-2005 average, respectively. By the end of this century, the mean sea level rise will approximately be 78 cm relative to 1989-2005 (Figure 3.1).

- Wind Speed: Computer simulations also predicted extreme wind speeds. It is expected that any predicted increase in wind speed might allude to the high probability of typhoons hitting or striking land in the future. For Hong Kong, the predicted extreme wind speeds are generally between 22%/yr and 1.4%/yr depending on the AR5 scenarios. For a 50-year return period, the mean hourly wind speed would increase by 2.11% by 2100 (Figure 3.2).

It was estimated that there would be an increase in construction costs for new coastal structures to cope with climate change. The increase in construction costs to design for climate change ranges from 1.0% to 3.9% of the capital construction costs. A benefit-cost analysis showed that the benefits brought by designing coastal

Sea surface height above Geiod

Representative concentration pathways (RCP) with radiative forcing value: +2.6, +4.5, +6.0, and +8.5 W/m²

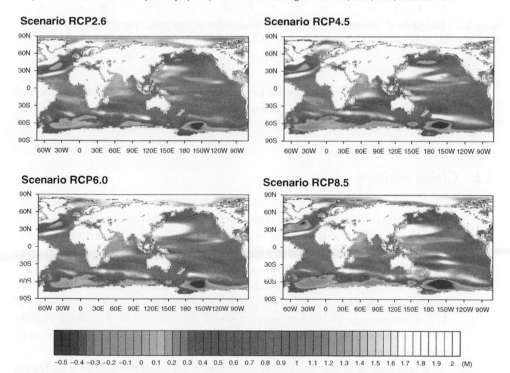

Figure 3.1 Projection of Sea-level Change at Year 2060 by CSIRO MK3 Climate System Model under AR5 Scenarios RCP 2.6, 4.5, 6.0 and 8.5. © Arup

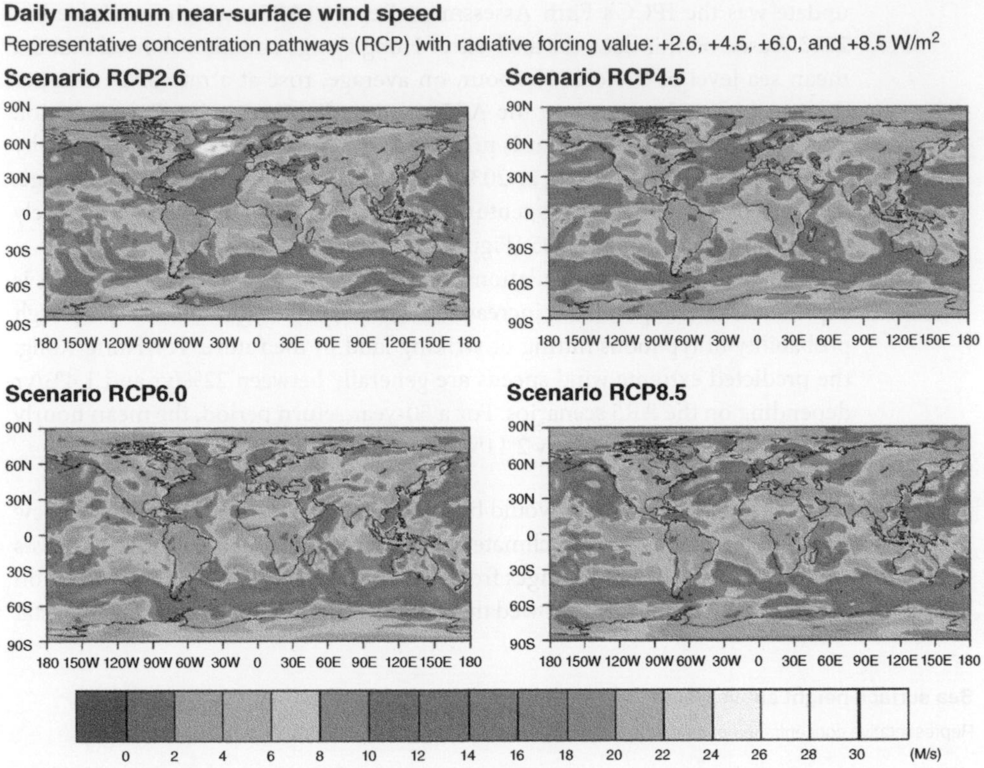

Daily maximum near-surface wind speed
Representative concentration pathways (RCP) with radiative forcing value: +2.6, +4.5, +6.0, and +8.5 W/m^2

Scenario RCP2.6   Scenario RCP4.5

Scenario RCP6.0   Scenario RCP8.5

Figure 3.2   Projection of Change of Daily Maximum Wind Speed at Year 2060 by Beijing Climate Center Climate System Model under AR5 Scenarios RCP 2.6, 4.5, 6.0 and 8.5. © Arup

defence structures such as seawalls for climate change far exceed the costs incurred by adopting such changes, and the ratios are at least 16% and 28% for design life of 25 years and 50 years respectively.

3.3.2   Global warming

Climate change could have a wide range of potential impacts such as global warming resulting in rising sea-levels, changes in precipitation patterns, and so on, that could profoundly affect all living organisms (including the human race) on the earth. In particular, the impact of a temperature rise would have significant impact on the living standards for humans. The Intergovernmental Panel on Climate Change (IPCC) indicated that the observed global average surface temperature exhibited an increase of 0.6 °C (0.4 °C to 0.8 °C) from 1901 to 2000. An increase of 1 °C of the average global temperature seems insignificant to the whole climate system. However, it is almost certain that seriously disruptive effects would occur with this seeming small increase in temperature.

It was indicated that the global atmospheric concentration of CO_2 increased from the pre-industrial level of about 280 ppm (parts per million) to 380 ppm in the year 2000, with a particularly sharp growth rate during the past 10 years

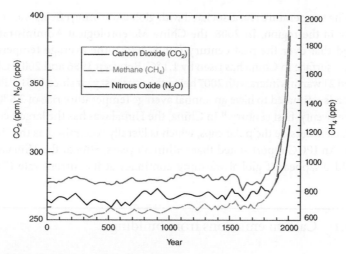

Figure 3.3   Build Up of Global Warming Pollutants. © Arup

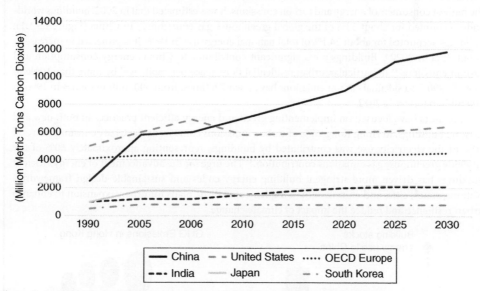

Figure 3.4   World Carbon Emissions by Region, Reference Case, 1990–2030. © Arup

(Figure 3.3). The concentration of CH_4 has increased from about 700 ppb (parts per billion) at the pre-industrial level to 900 ppb in the 1900s, and reached 1700 ppb in the early 1990, representing a growth rate of almost 90 ppb per decade. The global atmospheric N_2O concentration increased from a pre-industrial value of about 270 ppb to 320 ppb in 2005 (Figure 3.3).

In Asia, UN-Habitat research showed that more than half of the world's greenhouse gas emissions came from urban areas. Since 2006, China has become the world biggest CO_2 emitting country. Emissions from households, factories, vehicles and agriculture are irreversible if China's development model does not change its course in the coming years. By 2030, China's energy use and thus CO_2 emissions could double that of the United States (Figure 3.4).

The temperature rise associated with global warming effects was studied intensively in the region. In 2008, the China Meteorological Administration (CMA) stated that over the past century (1908-2007), the average temperature of the earth's surface in China has risen by 1.1 °C. Between 1986 and 2007, China experienced 21 warm winters with 2007 being the warmest year since 1951.[40] Furthermore, China is estimated to have an annual average temperature rise of 3.5 °C by the end of the twenty-first century.[41] In China, the Himalayas has the largest concentration of glaciers outside the polar caps, which is literally recognised as the "ice-tower" of Asia. An IPCC report stated that within 30 years, 80% of the Himalayan glaciers would disappear if global warming continues at its current rate (Example 3.1).

Example 3.1   Carbon emissions from buildings

The rapid pace of urbanisation and economic growth in Asian countries has resulted in an increase in demand for residential and non-residential building spaces. The building sector is identified as one of the highest consumers of energy and carbon emissions. It was estimated that in 2002, buildings worldwide accounted for about 33% of the global greenhouse gas emissions.[42] In China (Figure E3.1.1), buildings accounted for about 24.1% of total national energy use in 1996. It is projected to increase to about 35% in 2020.[43] Buildings are a significant contributor to China's energy consumption and carbon emissions.[44] In particular, urban residential floor space per capita will be more than triple to that of 1990.[45] Residential carbon emissions have risen 2.6 times, from 300 million tonnes in 1996 to 800 million tonnes in 2012.

Regulators have focused on implementing green and energy efficient practices in both new and existing buildings. In Hong Kong, 67% of carbon emissions come from electricity consumption and 90% of the electricity use was contributed by buildings, representing approximately 60% of the citywide greenhouse gas emissions contribution.[46] The urge for the advancement of energy efficient buildings has driven more stringent building energy codes and sustainable design frameworks. Nations, states, cities and citizens should implement realistic plans for urban expansion to enhance urban resilience and reduce the impact of climate change.

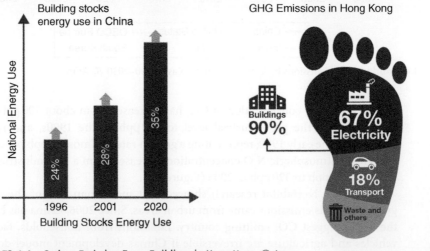

Figure E3.1.1   Carbon Emission From Buildings in Hong Kong. © Arup

Glacial melt will decline water resources and an increased variability of water would negatively affect almost half the human population and further exacerbate the serious water deficiency problem in China.

3.4   URBAN ENVIRONMENTAL DEGRADATION

Human disturbance causes environmental degradation, as earth's natural resources are depleted and the environment is compromised by the extinction of species, air pollution, water and soil pollution, and rapid population growth. Environmental degradation is one of the largest threats that are being closely examined in the world today. Urban living activities have intensified this process. This is particularly critical for regional developing countries as measures of control are not in place.

3.4.1   Air pollution

The rapid expansion of urbanisation has resulted in air pollution problems in cities. Air pollution in China is at an all-time high (Figure 3.5). The Asian Development Bank together with Tsinghua University in 2013 released the "National Environmental Analysis," which reported that among the world's 10 most air polluted cities, seven are in China. Those cities are the following: Taiyuan, Beijing, Urumqi, Lanzhou, Chongqing, Jinan, and Shijiazhuang.

It was reported that the average polluted days in Mainland China was almost 190 days in 2013, and 30 days were mainly affected by smog and PM2.5

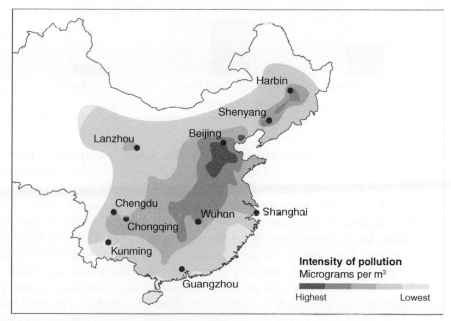

Figure 3.5   Air Pollution Problems in China. © Arup

Example 3.2   Cause of air pollution – vehicle transportation

Economic activity and inter-city migration has accelerated the load of domestic transportation within China. The number of passengers who travel by car and bus doubled from 800 billion to 2000 billion from 2003 to 2013 (Figure E3.2.1). The use of air transportation even tripled during this period and the demand for transportation is expected to grow.

Urbanisation in China has empowered the development of the transportation sector, in particular the automotive industry. Data published by the United Nations indicated that over 98% of the on-road vehicles in China are powered by fossil fuels such as oil and gas. Alternative fuels such as ethanol and natural gas products are hardly used as each source requires a unique distribution support system and requires expensive modification costs. No policy supports are found from The Central Government. The Ministry of Science and Technology in China has recently implemented the "863 Programme" and is committed to developing electric vehicles (EV) as a key form of transportation for the future. A detailed roadmap for EV has been laid down and enables a favourable policy environment to foster development.[47] It has been envisaged that electrified transportation could be an option for environmental friendly technology in the transportation sector if compared to other transportation energy options. However, the contribution of improving greenhouse gas emissions by pure battery powered electric vehicles would vary from region to region, which relies on the cleanliness of the source and the fuel-mix to power the electricity.

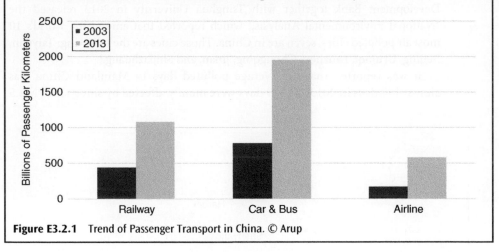

Figure E3.2.1   Trend of Passenger Transport in China. © Arup

particulates. Pollution caused by smog increased significantly during the winter months as coal was used as a major heating source; another major source was from the transportation sector (Example 3.2). More than 30 cities were severely affected by smog pollution, particularly Beijing and the Hebei region. The situation of smog can even be seen from space. Satellite images from NASA's Earth Observatory shows smog stretching 750 miles from Beijing to Shanghai. The satellite images showed naturally occurring white clouds along with smog appearing as grey swirls. When the photo was taken on December 7, 2012, harmful particles in Beijing were at 480 micrograms per cubic litre of air, which is almost 20 times the WHO's (World Health

Figure 3.6   Citizens Wearing Masks in Beijing. © Hung Chung Chih / Shutterstock.com

Organisation) guidelines. Due to the extremely high pollution levels, Beijing implemented a new plan to restrict construction and industrial activity, curb the use of vehicles used by government officials and limit outside activity for school children (Figure 3.6). The cost of pollution is measurable; in 2015, 700 flights were cancelled due to the low visibility caused by smog.

3.4.2   Energy depletion

The world's primary energy use increased from 10,198 million tonnes of standard coal equivalent (Mtce, 1 Mtce = 29.3 x 10^6 GJ) in 1980 to 17005 Mtce in 2006, representing an average annual increase of about 2% during the 27-year period. The total primary energy consumption of the world during 1980-2006 indicated a steady increasing trend of the world's total energy consumption. North America accounted for about one-third of the total energy used between 1980 and 2006. During this same period, a modest increasing trend in primary energy use was observed for Central and South America, Europe, Middle East, and Africa. There was a slight fluctuation in Eurasia. The energy use in Asia and Oceania exhibited a continuous increasing trend. It overtook Europe in 1991 and North America in 2003, which was mainly due to the robust economic growth in a number of Asian countries, especially China (Figure 3.7).

The U.S. Energy Information Administration (EIA) presented a projection of the world's marketed energy use outlook until 2040.[48] It was expected that fossil fuels would continue to be the largest source of energy supply worldwide. The oil share of

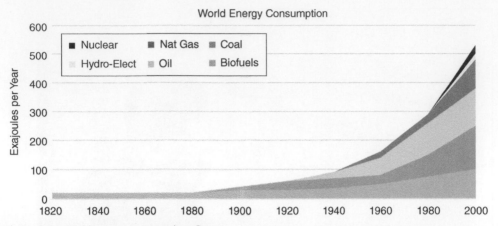

Figure 3.7   World Energy Consumption. © Arup

world marketed energy consumption was predicted to fall from 33% in 2012 to 30% in 2040, due to the projected high oil prices and the shift to other energy sources. Natural gas remains as an important fuel for generating electricity worldwide. According to the U.S. EIA, the world's total natural gas use is expected to increase by 1.9% per year on average, from 120 trillion ft³ in 2012 to 203 trillion ft³ in 2040. The high global oil prices encourage consumers to turn to natural gas, but technical difficulties on supply will slow down the growth of natural gas after 2020. In general, it was envisaged that renewable energy sources and nuclear power will become an up-and-coming energy source. Worldwide coal consumption increased significantly between 2001 and 2009, largely due to the growth in the coal consumption of China (from 24% to 29%). Between 2007 and 2012, however, coal consumption declined by OECD due to the global recession. The world's coal consumption was projected to increase by 0.6% per year on average from 2012 to 2040, with the peak growth in demand will occur around 2020.

The demand for energy consumption will be accompanied by the mushroom effect of economic growth. Urbanisation has driven up the demand of energy for most Asian countries, and the quest for energy has become a geo-political issue in many parts of Asia. The national total energy consumption of China's total energy use rose from 603 Mtce in 1980 to 2850 Mtce in 2008, representing an average annual increase of 5.7% during that 29-year period. Coal accounted for about 72.2% and 68.7% of the primary energy consumption in 1980 and 2008, respectively. During the same period, petroleum varied from 20.7% to 18.7%, hydroelectric power 4% to 8.9%, and natural gas 3.1% to 3.8%.

Since 2009, China's coal imports have been greater than coal exports. Thus, the government is faced with the need to develop alternative energy systems. The policies of the Chinese government are clearly directed towards a greater thrust on energy and environment under the framework of the National Five-Year-Plan. Conservation policy and energy efficiency investment have continuously been made in the last two decades. The twelfth Five-Year Plan in 2011 targeted to reduce the energy intensity by 16% and carbon intensity (i.e., the amount of GHGs or carbon emissions per unit of GDP) by 17% (Example 3.3). Likewise, other regional countries are also striking for more non-fossil fuel application (Figure 3.8).

Example 3.3   Creation of renewable market in Asia

China's Non-Fossil Energy Development
Achieving a high proportion of renewable energy has become a major national energy strategy of China. As proposed in its Intended Nationally Determined Contributions (INDC), China intends to raise the proportion of non-fossil energy in primary energy consumption to about 20% by 2030. That ambitious goal means that non-fossil energy supplies by 2030 will be 7-8 times that of 2005, and the annual increase rate will be more than 8% within 25 years. Besides, the capacity of wind power, solar power, hydropower and nuclear power will reach 400 GW, 350 GW, 450 GW, and 150 GW respectively, and China's non-fossil power capacity will be greater than the US's total power capacity. In addition, the scale of natural gas will also increase. Consequently, by 2030, the proportion of coal will fall from the current 70% to below 50%, and the CO_2 intensity of energy consumption will decrease by 20% compared to the level of 2005, which play important roles in significantly reducing the CO_2 intensity of GDP.[49]

Japan Sets 22-24% Renewables Share Target for 2030
The recent incident of the Fukushima nuclear power plant has triggered the discussion of wider use of renewable energy in Japan. Japan's Ministry of Economy, Trade and Industry (METI) has recently formulated energy plans with an aim to increase the renewable portions in the national energy supply to 22% to 24% of the country's power mix by 2030; and this will substantially decrease the share of fossil fuel usage. In Japan, solar power is the top renewable energy source. To meet this 2030 target, Japan needs to install photovoltaic (PV) capacity of 64 GW.

Korea's Fourth Basic Plan for New and Renewable Energy
The government of South Korea announced a plan in September 2014 to identify and set a specific target of providing 11% of the country's total primary energy supply with new and renewable energy by 2035. It begins with increasing the mean annual growth rate of renewables to 6.2% from 2014 to 2035 from the current 0.7%. The plan seeks to invest in energy efficient buildings (including residential buildings) in order to reduce energy waste. The blueprint focuses on expanding public and private partnerships in a new and renewable energy sector. This will include building a market base for renewable energy and reducing an over-emphasis on government-led financing of clean energy projects.[50]

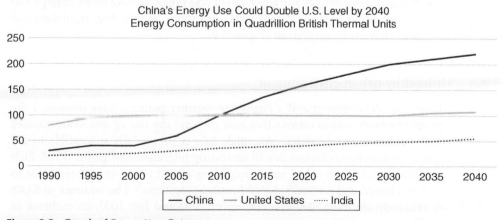

Figure 3.8   Trends of Energy Use. © Arup

3.4.3   Waste generation

The modern ways of urban living have created the problem of excessive waste for cities in Asia. A survey released by the Environmental Protection Department of Hong Kong SAR indicated that domestic waste accounts for the major portion of total waste generation of the city, followed by construction waste. In China, data from the National Bureau of Statistics revealed that the national MSW was 30 million tonnes in 1980 and rose to 170 million tonnes in 2013, or 5.5-fold over that period. The amount of MSW treated significantly increased in 1994, or 30% of its total volume. In 2013, the rate of MSW treated reached 90%. There are three major types of MSW disposal treatments, namely landfill, composting, and incineration. The waste treatment strategy is becoming an issue in society as some methods such as incineration and landfills are controversial and may jeopardise public health in the long run. The treatment capacity of MSW disposal by incineration was 3.7 million tonnes per year in 2003, or 4.9% of 3 types of treatment. In 2013, the MSW treatment capacity by incineration increased to 46.3 million tonnes, representing 30.1% of the total treatment. The increase was 12.5 times over a 10-year period.

Minimising the environmental impact of waste management is key to sustainable use of the ecological environment. Recovery is one aspect of sustainable waste management that is based on the well-known hierarchy of prevention, reuse, recycling, recovery, and disposal.[51] Waste-to-energy (WTE) is a kind of sustainable waste management, which turns non-recyclable MSW into energy. It refers to the recovery of heat and power from waste, and in particular non-recyclable waste.[52] However, people's poor understanding, doubt of safety and weak investment may set difficulties of the WTE technology and introduction in the region and China in particular.

The experience of some cities, such as Taipei and Seoul, has suggested the importance of reducing waste generation at its source. The concept of "polluters pay" is strictly implemented and resulted in much lower levels of domestic waste generation. On the other hand, the polluter pays concept or principle might be too pre-mature for other developing countries to implement. There are a number of obstacles that may occur during the implementation process. At a rationale level, incentives of polluters pay the government and polluters have to set examples to the society. At an implementation level, operations, monitoring, regulation and execution have to be carried out (Figure 3.9).

3.4.4   Unhealthy urban environment

Economic development and a change in weather patterns have promoted the people mobility within regions that may increase the rise of infectious diseases with epidemic and pandemic potential. Asia is particularly susceptible to the spread of infectious diseases due to increasing migration and global travel, high population density in urban areas, increasing spread of tropical diseases due to climate change, and underdeveloped healthcare systems.[53] The outbreak of SARS is extremely alarming. Between November 2002 and July 2003, an outbreak of SARS in southern China caused an eventual 8,096 cases and 774 deaths reported

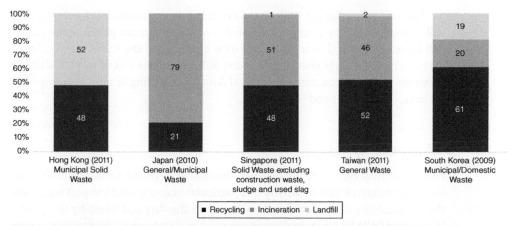

Figure 3.9   Comparison of Waste Management Structure with Other Areas in Asia. © Arup

in multiple countries with the majority of cases in Hong Kong (9.6% fatality rate) according to the World Health Organisation (WHO). Within weeks, SARS spread from Hong Kong, infecting individuals in 37 countries in early 2003. Many people were concerned about the poor quality of the urban living environment, which led to the outbreak and rapid spread of the disease in several building estates in Hong Kong, and the lack of preparedness of the medical system for pandemic and epidemic diseases. Regional collaboration is crucial to contain the risk of spread and handle the situation if an outbreak occurs.

3.5   LIVEABILITY DEGRADATION

Urban environment defines the liveability of a city, and to a certain extent on how we perceive the quality of life. Some cities are poorly planned and result in unfavourable conditions for living and working, particularly for high-density cities, where the right to air and light is not protected. Many urban issues such as urban heat island and ecological degradation are prevailing. These urban design problems are severely affecting the health and well-being of the city dwellers.

3.5.1   Urban heat Island

The urban fabric has changed the capacity of urban environments to store more heat during the day than that of rural areas. This stored heat is then released at night and causes night-time air temperatures to be higher in urban areas than the rural areas. Urban Heat Island (UHI) is defined as the temperature difference between urban and rural areas. Urban fabric, characterised by the reduction of greenery and impervious surfaces causes UHI. The problems of UHI are far reaching. Elevated temperature causes heat stress, which may potentially affect the health and wellbeing of city dwellers, particularly the young and elderly. It also causes a surge in energy usage in buildings as more energy is used for air conditioning systems.

The long-term effects of UHI are unknown but the occurrence of global warming will exacerbate its implications. It is predicted that the mean global temperature will increase by 2-4 °C over the twenty-first century and the surface temperature may rise by 3.5-5 °C. Higher temperatures will expand the risk of tropical disease occurrences. In tropical and sub-tropical Asia, the breeding of mosquitoes spread malaria, dengue fever and filariasis.

3.5.2   Ecological footprint

Human activity can affect the delicate eco-system of planet earth. An ecological footprint measures human activity in a particular country, which in part measures the sustainability of that country. According to the data published by the World Wide Fund (WWF), China and other countries in Asia have an observable increase in the human development index associated with urbanisation. This also resulted in an increase of their ecological footprint in different degrees. In general, developed cities in Asia have already moved outside the sustainable quatrain (Figure 3.10). It will take extra effort for these cities to reverse the course of their development. The development in China is of particular concern, due to its immense impact on the world. Ecological footprint of China is now on the boundary of sustainable development quadrant. It is critical for China to formulate a policy that will change the resource-intensive development to a more sustainable mode of development, not following in the footsteps of other developed countries.

As indicted in Chapter 2, the construction industry and the built environment are the two key areas that must change in order to achieve sustainable development in our society. The current consumption of regional developed countries is far beyond what planet earth can sustain. Building green and living green are the only viable solutions for regional developing countries to reduce their ecological footprint and bring the trajectory of their development curve into the Global Sustainable Development Quadrant.

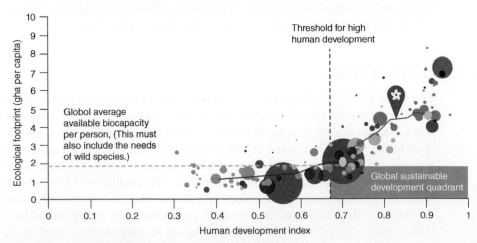

Figure 3.10   Ecological Footprint of Countries. © Arup/WWF

3.6   SUMMARY

The unprecedented urbanisation in Asia over recent decades has stretched the carrying capacity of planet earth to its limit. Global carbon emissions have reached a level that has caused climate change or climate extreme, affecting every corner of the world. Extreme weather poses a significant threat to our lives and property. Asian countries have no exceptions. The situation is more vulnerable for developing countries, such as China, where the intensive process of urbanisation has imposed severe environmental loads and resulted in environmental problems, such as poor air quality affecting the health of people. Building a city's resilience to climate change is a priority for many policy leaders in the region. For those regional developed countries, the negative effects of urbanisation are also challenging. These countries are consuming more and more resources – energy in particular – to improve people's quality of life and to sustain their reckless consumption behaviour. This has resulted in a vicious cycle of uncontrolled carbon emissions. Unless a sustainable green movement is put in place to rectify the situation, natural disasters and environmental catastrophes are expected to occur at an increasing rate in the future.

Chapter 4
Quest for solutions

4.1 INTRODUCTION

Various parties worldwide from international organisations, governments, business leaders and professionals are looking for solutions for the problem of climate change and urban challenges. As a result of global collaborations in the past few years, a consensus was formed that sustainable development is the ultimate solution to handle the challenges that is facing our planet and our living environment. There are many leading forces from the stakeholders pushing in this direction, namely (1) governments have provided incentives to combat environmental issues, (2) the business sector is set to increase market competitiveness, and (3) the individuals are in pursue of a better quality of life. It is a win-win solution for all. Yet, how to deliver a plan still remains as a key question.

International collaborations have played an important role in addressing this global issue in recent years. At an inter-governmental level, the United Nations is taking the lead in studying the implications of climate change under different scenarios of carbon emissions. Targets on carbon emission reduction for the developed and developing countries have been agreed under the Kyoto protocol and the recent Paris COP21 Agreement. With these framework agreements in place, all nations are looking for solutions that can resolve not only their own environmental problems but also the global challenge of climate change. At a country level, non-governmental organisations are also playing an active role in aligning the resources from civic and business societies to achieve the goal of low carbon development.

The realisation of our common goal of sustainable development requires a concerted effort from all stakeholders, namely governments, practitioners and the public. Their roles and responsibilities need to be defined clearly. For less developed Asian countries in particular, institutional arrangement is lacking in implementing new strategies. A partnership with the building industry can be a viable option to provide the necessary resources for driving the green movement.

Building Sustainability in East Asia: Policy, Design, and People, First Edition. Vincent S Cheng and Jimmy C Tong.
© 2017 John Wiley & Sons Ltd. Published 2017 by John Wiley & Sons Ltd.

4.2   HISTORY OF INTERNATIONAL COLLABORATIONS AND PARTNERSHIPS

Over the past few decades, many collaborations and partnerships happened on international and intergovernmental levels, leading to today's global actions in addressing climate change. Some references were available on the detailed accounts of these events.[54,55] The following paragraphs summarise the history of key events in chronological order:

In 1972, the UN conference on Human Development was held in Stockholm. The topic was on pollution and acid rain in northern Europe. The conference led to the development of the United Nations Environment Programme (UNEP).

In 1979, the World Meteorological Organisation (WMO) organised the first World Climate Conference. The concern of regional and global climate change was raised and called for global cooperation to explore the possible future course of global climate.

In 1985 a joint UNEP/WMO/ICSU Conference was held in Austria, based on the impact of a build-up of carbon dioxide and other greenhouse gases. It concluded with a prediction of global warming.

In 1987, the tenth Congress of the WMO was organised and recommended UNEP to establish an intergovernmental mechanism to study climate change. It led to the formation of the Intergovernmental Panel on Climate Change (IPCC) in 1988. Separately, the Brundtland Report that weaves together social, economic, cultural and environmental issues and global solutions was released. It popularised the term "Sustainable Development".

In 1992, the United Nations Framework Convention on Climate Change (UNFCCC) was formed. It was signed at the United Nations Conference on Environment and Development in Rio de Janeiro, also known as the "Rio Earth Summit".

In 1997, the Kyoto protocol was finalised. It was the first agreement between developed countries to mandate reductions in greenhouse-gas emissions. Nearly all developed countries ratified the treaty in 2005, with the notable exception of the US.

In 2015, the Paris COP21 Agreement was signed, further incorporated the developing countries into formulating an action plan. The world now has a common goal of controlling carbon emission levels.

Since rectifying the Kyoto protocol, many collaborations on intergovernmental, national, and city level governments as well as the business sector were conducted to iron out the details of implementing sustainable development in the region. These experiences have demonstrated the importance of engaging all the stakeholders for a concerted effort to address the issues of climate change and urban challenges. In some cases, the private sector initiatives proved to be very effective in the region. C40[56] and the World Economic Forum Partnerships programs[57,58] are amongst the successful cases.

4.3   C40 CITIES CLIMATE LEADERSHIP GROUP INITIATIVE

As highlighted in Chapter 3, many cities are facing a lot of urban challenges that are jeopardising the well-being of people and social support systems. In principle, many actions for addressing these challenges can be taken at a city level. To explore the capacity of megacities for addressing climate change, the C40 Cities Climate Leadership Group (C40) was formed, and it works as a network of the world's megacities to reduce greenhouse gas emissions.[59] Harnessing the assets of member cities, C40 forges to address climate risks and impacts locally and globally. The 40 participating megacities include London, New York, Beijing, Shanghai, Tokyo, Seoul and Hong Kong. In specific, the East Asia cities represent 297 million people and generate 18% of global GDP and 10% of global carbon emissions. Collectively, mayors of the C40 cities are capable of creating a huge potential impact by implementing climate change actions.

In 2011, the C40 commissioned Arup to carry out a comprehensive survey for the mayors of C40 cities on the actions undertaken by cities to address climate change. The aim of the survey was to understand the powers, actions and opportunities of the 40 members to reduce carbon emissions and adapt to global warming. The 2011 report (Version 1) demonstrated that C40 cities have many of the powers necessary to mitigate climate change, and adapt to it. It also demonstrated that the mayors of C40 cities are already using their power to take action on reducing greenhouse gas emissions. In 2014, the C40 membership expanded to 63 cities, and the survey report (Version 2) found that the C40 network was an effective way of spreading and accelerating best practices. The date further shows that "…developing cities' best practices are being adopted by developed cities as much as the other way round."[60] During a COP21 meeting in Paris in December 2015, C40 released a third report, CAM 3.0, which stated that a potential annual saving of 645 Mtonnes CO_2e by 2020 can be achieved through the climate actions by member cities (Figure 4.1).

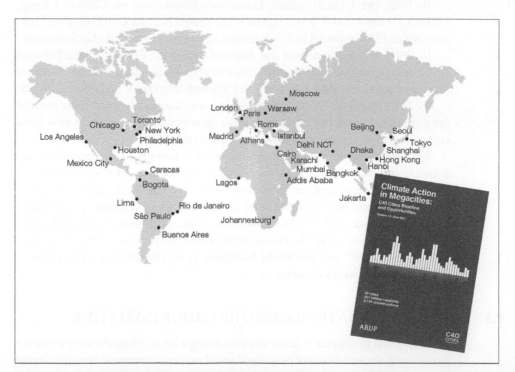

Figure 4.1   C40 Participating Cities and the Actions Taken by Their Mayors. © Arup

4.3.1   Key issues

Many national governments and intergovernmental organisations are working on the actions necessary to prevent catastrophic climate change. Not all problems of global warming can be tackled on a city scale. In particular, national governments have to take responsibility for large-scale infrastructure and social support facilities. Notwithstanding, this study demonstrates that C40 city mayors have the capacity and authority to dramatically reduce carbon emissions in their cities. In particular, C40 cities have the power to take action in nine sectors, namely transport, existing buildings, waste management, water, energy supplies, outdoor lighting, planning and urban land use, food sources, and agriculture. The study demonstrates that the C40 is making a significant difference in improving the scale and speed for tackling climate change. More than 9,831 climate change actions have been in effect since 2011. It also shows that cities have also taken initiatives to transform policies and infrastructure to improve energy efficiency and saving of other resources. The C40 cities are expanding their capacity in taking on more actions. Most importantly, C40 networking can help to facilitate the transfer of knowledge and the scaling of further action.

4.3.2   Action plan on buildings

On average, buildings account for 45% of C40 cities' carbon emissions. However, this proportion varies considerably across members' cities. For example, in New York the figure is 78% compared with just 12% in São Paulo. Climate (New York has both very cold winters and hot summers while São Paulo's climate is relatively moderate), cultural differences, and occupancy levels are factors that affect the energy consumption levels of buildings.

Existing buildings are a major concern. In many highly developed C40 cities it is estimated that more than half of all existing buildings will still be in use in 2030. Considerable focus is therefore in need to devise emission reduction activities on reducing the energy demand of existing buildings in these cities. "Retrofitting" buildings (refurbishing) is a possible solution. It is a more cost-effective way of improving energy efficiency, rather than tearing down and re-building. The report found that reducing energy use in buildings can have a multiplier effect; every unit of electricity saved in the home or office can translate into three units saved at the power plant. This is due to the inefficiencies of transmission and distribution in our ageing energy infrastructure.

Overall, mayoral powers in the building sector are strong among the C40 cities because more than half of city governments own or have power over public sector building stock, including municipally owned offices, public sector housing, and other facilities. Some mayors indicated that initiatives and actions in the private sector are also within the power of some city governments. It is notable that while mayoral powers of ownership, operation, budget control, levy setting, and vision setting vary considerably across different building types, mayoral powers to set or enforce policies and regulations are consistently fairly strong in the building sector.

4.4   WEF PARTNERSHIP FOR FUTURE OF URBAN DEVELOPMENT

On addressing climate change, partnership with the private sector forms a key part of the solution. In particular, the private sector is good at innovation, and pioneering the investments of appropriate technologies in order to facilitate the development of capacities.

The World Economic Forum (WEF) is one of the leading global organisations that champion sustainable development in the business sector. The WEF observed that most cities in developing countries lack the capacity and resources for sustainable growth. It is believed that through partnerships with businesses, "… Cities have the opportunity to rely on new business models and technology solutions to enhance their liveability, sustainability, efficiency and productivity" and "… Cities are driving economic development and fostering innovation while playing an important role in reducing poverty and addressing climate change."

Since 2014, WEF has initiated shaping the Future of Urban Development & Services (FUD) initiative with the aim to be a partner in the transformation of cities around the world. FUD creates a platform and links city governments with business leaders in partnerships to resolve the issues of urbanisation. Since its inception, FUD has developed a strong brand and is recognised by the partnering cities. In its first two terms (2012-2014), FUD worked with the Champion Cities of Tianjin, Dalian and Zhangjiakou of China, to address their most pressing urban development challenges. In particular, the FUD initiative selected the following three urbanisation action themes:

1 Transport planning and management
2 Urban energy management
3 Sustainable industry development

For each of these cities, an Urbanisation Issues Framework was developed based on these three themes to map the problems these cities currently face. Six key issues were then identified by the initiative to provide strategic recommendations (Figure 4.2). In operation, these strategic recommendations were stemmed from the international and Chinese experts of the community of the FUD initiative and from both the public and private sector. They provided the best practices and insights on policy frameworks and various business models.

As concluded by the director of Arup who is one of the partnering companies, the implications of the FUD program is crucial for local governments to realise sustainable development:

> "Arup is very pleased to contribute to the initiative as the Project Champion. Tianjin is one of fastest-growing cities in China, and the pace and scale of growth have also brought issues and challenges to the city's long-term development. I am particularly impressed by the openness and insightful views offered by the global experts, industry leaders and local collaborators, and together we came up with strategic recommendations to address the urban goals of Tianjin. We believe these recommendations, comprised of both near-term and long-term strategies, provide a good base for implementation as we move from vision to action" – *Michael Kwok, Arup, a partner of FUD program.*

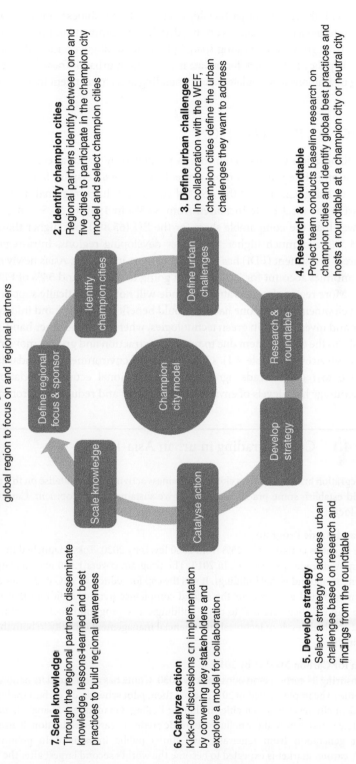

1. Define regional focus & sponsor
The initiative's steering board identifies a global region to focus on and regional partners

2. Identify champion cities
Regional partners identify between one and five cities to participate in the champion city model and select champion cities

3. Define urban challenges
In collaboration with the WEF, champion cities define the urban challenges they want to address

4. Research & roundtable
Project team conducts baseline research on champion cities and identify global best practices and hosts a roundtable at a champion city or neutral city

5. Develop strategy
Select a strategy to address urban challenges based on research and findings from the roundtable

6. Catalyze action
Kick-off discussions on implementation by convening key stakeholders and explore a model for collaboration

7. Scale knowledge
Through the regional partners, disseminate knowledge, lessons-learned and best practices to build regional awareness

Labels within the figure: Champion city model; Define regional focus & sponsor; Identify champion cities; Define urban challenges; Research & roundtable; Develop strategy; Catalyse action; Scale knowledge

Figure 4.2 The FUD Approaches for Champion Cities. © Arup/WEF

With the success of partnerships between three Chinese cities, FUD turned their focus to India in 2015 and recommended three recommendations for India to facilitate inclusive growth: integrating spatial planning at all governmental levels, creating a stable policy framework for private investment in urban infrastructure and formulating institutions to stimulate capacity building and attract talent to grow businesses.

4.5   REGIONAL INTEGRATION

Collaboration of Asian countries is propelled by regional economic integration. Economic integration is most evident in East Asia (Japan, South Korea, Taiwan, Hong Kong, China and the ASEAN countries). Intra-regional trade as a share of East Asia's total trade increased from 36.8% in 1980 to 54.5% in 2006.[61] That is lower than the comparable share for the EU (65.8%), but higher than for NAFTA (44.3%) and much higher than other developing regions. Intra-regional Foreign Direct Investment (FDI) has also become more important. Asia's newly industrialised economies account for 29.2% of FDI going to ASEAN and 54% of FDI in China.

More regional cooperation on trade will not automatically support sustainable development. On the one hand, it would benefit technology and information transfer and investments in green technologies, whereas on the other hand, it may create risks to the environment due to resource extraction and greater movement of goods and services (Example 4.1). Convergence of environmental standards must be in place to regulate trade agreements, and regional economic relationships must encourage high levels of environmental quality and reduced environmental risk.[62]

Example 4.1   Carbon trading in urban Asia-Pacific

Regional integration brings countries closer for business activities. To capitalise on this network, the region should establish some practices/rules to drive sustainable development. Carbon trading is one topic of focus.

Tokyo Cap-and-Trade Program
With an aim to cut its emissions by 25% from 2000 levels by 2020, Tokyo launched the world's first city-level "cap-and-trade" programme in 2010. The program covers 1,340 existing large commercial, government, and industrial buildings. It sets the cap for reducing GHG emissions at 6% for the first compliance period and 15% for the second compliance period. Buildings that reduce more than the cap can sell their "credits" to other buildings. The program benefits from a cooperative environment and high levels of technical and financial management capacity in both the public and private sectors.[63]

China's Plan for Carbon Market by 2017
Aiming at ensuring its carbon emissions peak by 2030, China has announced its national emissions trading scheme to be implemented in 2017. Before then, pilot schemes on carbon trading have been implemented in city-level in seven cities including Beijing, Guangdong, Shanghai, and Shenzhen since 2013. The markets now allow polluters to trade credits in carbon reduction. It also encourages more power generation from renewable and limits public investment in polluting projects. Guangdong's carbon market is expected to become the world's second largest after the EU.

4.6 CHANGES FOR SOLUTIONS

Driven by the pressing situation of climate change, all countries worldwide are searching for viable and lasting solutions. Through global collaborations from various organisations in the past few years, the world has come to the conclusion that we all need to change our course of resource-intensive development into sustainable development. The problems created by the process of urbanisation in the past few centuries cannot be resolved by any simple solution. In fact, the issues of climate change are multi-faceted on various social, economic and environmental issues, and the solutions are bound to be multi-dimensional involving government policy, partnership with business sector and behavioural changes in individuals. They have to be dealt with at different levels, from the state level down to the individual citizen level. In the past, actions for carbon emission reduction have been considered across all sectors, including the industry, transportation, buildings, and so on. Experiences of western countries in the past two decades have verified that measures in the building sector are by far the most effective, and therefore, receive the highest priority in their policy.[64] This lesson is particularly crucial for Asia, where a lot of construction is happening to support the needs of rapid urbanisation. It is imperative that we focus on buildings and discuss the changes in the stakeholders for the solutions for addressing one big question – how can we deliver quality living space for people whilst minimising carbon emissions?

4.6.1 Re-think of sustainable development framework

Delivering sustainable development is a complicated process and requires various levels of involvement/engagement from stakeholders. At a state level, the government needs to establish an appropriate policy to achieve the national target of carbon reduction. At a policy implementation level, the industry needs to develop the practices for solving technical and financial matters. Standardisation of design and practice plays an important role to build the capacity of a green movement within society. On an individual level, people must change their behaviour to embrace low carbon living. These three parties, that is, government, practitioners and the public are highly interconnected when making their decisions towards a green movement. For example, government policies will have a strong influence on industry practices. Practitioners will expect the government to set a bottom line for operations by requiring an entire industry to adhere to minimum standards on controlling issues. Likewise, people will have an expectation of an improved living environment and will demand that the government and the industry meet the requirements (Figure 4.3).

There will be conflicts on implementing sustainable development whenever the rights of any stakeholders are infringed upon or the benefits are not fairly distributed. The government plays a pivotal role in resolving conflicts and aligning all parties for compromised solutions. In the past, this always happened in redevelopment projects in China where land resumption occurs. The government has the indisputable role to resolve the conflicts. In Singapore, we also see some good signs of strong leadership in the government on engaging its people on issues of controversy to arrive to a consensus. Without public support, sustainable development will never be possible.

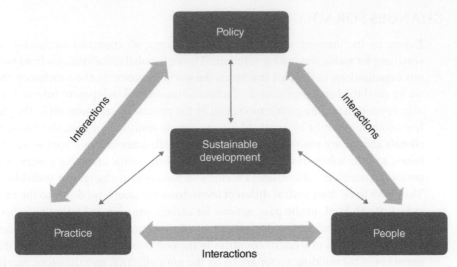

Figure 4.3   Framework for Realising Sustainable Development. © Arup

4.6.2   Issues of policy

After the Paris COP21 agreement, all countries, developing and developed alike, will have to make a commitment to their action plan for carbon emission reduction. Transformation into a sustainable development model is the only solution for all countries. To achieve this goal, the government has a crucial role in formulating its policy to direct all the necessary focus and resources from society. Recent experiences of Europe provide some good references. In the United Kingdom (UK), the Climate Change Act of 2008 was an act of Parliament. The Act established a legally binding target to reduce the UK's greenhouse gas emissions by at least 80% below base year levels by 2050.[65] On implementation, the Act introduced a system of carbon budgets, which is a legally binding limit of carbon emissions produced in successive five-year periods, beginning in 2008. The first three carbon budgets were set into law in 2009, requiring emissions to be reduced by at least 34% below base year levels in 2020. The fourth carbon budget, covering the period 2023-2027, was set into law in 2011, requiring emissions to be reduced by 50% below 1990 levels. Asian countries followed the same practice, first in Japan then in Korea, Singapore and China and India.

The government also has a role of levelling the playing field of industries by stipulating regulations to regulate business operations. Building energy efficiency is a good example of this. In most urbanised Asian cities, buildings constitute 60-70% of total energy end use and approximately 90% of electricity consumption. Energy saving opportunities is huge on the energy efficiency of buildings. In Hong Kong, Energy Efficiency Regulations for buildings came into operation in 1995, which was the first set of legislation in Hong Kong to control energy-efficiency design in buildings. It specifies statutory control on the design of building envelopes of new commercial and hotel buildings by using the Overall Thermal Transfer Value (OTTV) method. Four sets of energy efficiency codes of practices for electrical and mechanical systems have been designed and implemented voluntarily since 1998.

To promote the energy efficiency of buildings, the Hong Kong Government enacted the Buildings Energy Efficiency Ordinance, which came into full operation in September 2012. Since then, all buildings are required to meet the energy codes consistently.

4.6.3   Issues of practice/design

Sustainable development is a new concept for the building and construction industry. Although the idea of a low carbon city or buildings has been promoted extensively in recent years, the practitioners (be it architects or engineers) have difficulty putting the idea into practice consistently. There is no one single design standard on sustainable development being practiced worldwide. Different codes have different focuses and criteria to define sustainable development. Many issues, in particular social and economic ones, are difficult if not impossible to quantify and measure. There has been significant progress in the design of green buildings in recent years. Thanks to the collaboration of non-governmental organisations of the green buildings community and professional institutions worldwide, the key principles of sustainable buildings such as life-cycle approach and integrated design are becoming mainstream for the government projects. The standardisation process of green practice is ongoing within the construction industry.

In the market, adoption of sustainable development is still at an initial stage. Its wider adoption needs incentives from both the government and the market itself. For instance, the Building & Construction Authority (BCA) of Singapore has launched the second Green Building Master plan to set out specific initiatives to achieve a truly sustainable environment. To encourage the private sector to engage in this sustainable building development, the Green Mark (GM) GFA Incentive Schemes have been implemented. Buildings that achieve GM highest Platinum rating can qualify up to 2% additional GFA (with a cap of 5,000 m²), whereas the second best Goldplus can qualify up to 1% of additional GFA (with a cap of 2,500 m²). In China, the Twelfth Five-Year Plan aims to reduce overall energy use by 16% per unit of GDP equivalent to reduction of CO_2 emissions by 17% per unit of GDP by 2015. In China's Ministry of Finance and the Ministry of Housing and Urban-Rural Development committed to 30% of all new construction projects achieving China's National Green Building Evaluation Label Three Star certification by 2020. A subsidy of ~13 US$/m² will be awarded for the highest rank three-star rating building and ~7 US$/m² for the second best two-star certified buildings. The monetary incentive schemes are recognised as the largest motivation for sustainable development, which offers direct financial links and drives the overall green building market in China. These incentives help to build the capacity of the practitioners for implementing sustainable development.

4.6.4   Issues of people

The main objective of implementing sustainable development is for the benefit of the people. Many governments and the professionals have looked into the problems of our built environment, which are affecting the health and well-being of people. In many Asian cities, the issue of enhancing "liveability" is crucial because of the unique high-rise and high-density living environments. Designers are applying technologies and innovations to create quality space for people living in these compact areas. In Hong Kong and Singapore, the governments are applying more resources to create a sustainable community in public housing developments. These provisions are proving to be essential to fostering a sense of belonging for its occupants.

Ironically, people are also one of the main hindrances in achieving sustainable development. Many studies have found that energy and other resources of a building are wasted on the reckless behaviour of occupants. Our daily patterns of consumption at home and work are the cause of many environmental problems. Advocating sustainable (or low carbon) living is as important as making our buildings or infrastructure more efficient. Educating the public, government influence, and the business sector are critical to initiate and sustain the changes in behaviour of individuals, in particular, their lifestyle and consumption habits.

4.7   PARADIGM SHIFT

The top-down approach is prevailing on the implementation of green buildings worldwide, whereby the government lays down a policy and formulates regulations for industries to follow. At an early stage, this kind of one-way approach is effective as it can ensure a consistent implementation. Yet, the market force has not been fully capitalised and people (or stakeholders) are not entirely engaged. The momentum needs to re-align for concerted efforts.

This book has been written considering these layers of thought, starting with policy, design and people, and finally the market, to transform the current practice and reverse the course (or the mode in specific) of the urbanisation process. The next few chapters describe how the ecological age of a built environment can happen through a joint effort by governments, various stakeholders, and advanced engineering approaches.

On policy

Section 2
un policy

Chapter 5
Policy framework

5.1 INTRODUCTION

The world has witnessed the largest global collaboration on combating global warming due to the excessive carbon emissions associated with human activity. Many governments have made commitments on emission reduction and have taken the necessary steps to address the problems. At the government level, policies have been formulated to change the operations of industry, business and individuals to a more sustainable manner. Developing a low carbon society/economy is fast becoming the prevailing initiative worldwide.

Asia is no exception in advocating this green movement. For the more developed East Asian economies such as Japan, Korea, Singapore, Hong Kong, and Taiwan, sustainable development is considered the key subject on the national agenda, not only for the sake of enhancing environmental performance but also as a strategy for enhancing social and economic benefits. Policies on directing industry to practice sustainability are in place to increase competitive advantages in the global arena. Developing or emerging countries, such as China, treat this as imperative for the survival of the nation during their process of urbanisation. Policies have been channelled to focus on the solutions and strategies that can help resolve the environmental crisis.

In the quest for solutions, government policy plays a pivotal role, channelling limited resources to address the most pressing issues. Adopting the appropriate measures and institutional arrangements are critical to the execution of the policies. The success of the green movement will hinge on the prioritisation of policies and their effective implementation. This chapter will develop a model to study the framework of the sustainability policies of the building sector that prevail in Asian countries, highlighting the key subjects on the agenda and the effectiveness of these policies.

Building Sustainability in East Asia: Policy, Design, and People, First Edition. Vincent S Cheng and Jimmy C Tong.
© 2017 John Wiley & Sons Ltd. Published 2017 by John Wiley & Sons Ltd.

5.2   POLICY FRAMEWORK

Due to the rapid change of the built environment in Asian cities (resulting from urbanisation) and global concerns on climate change, there is always a case for reviewing the policies that affect the building and construction industry in Asia. They are broadly reviewed as being on the agenda of sustainability. The agenda includes carbon emissions, energy consumption, waste management, and air quality, and so on (refer to Chapter 3 on the background of these imperatives). In recent years, many governments have adopted the requirement for new regulations and that a Regulatory Impact Assessment (RIA) should be undertaken.[66,67] A RIA typically consists of identification and clear statement of purpose for the regulations, their risks, fairness, and benefits, which, if possible, should be given a value. However, the main issue for the RIA is in assessing the costs that are imposed on society as well as any other indirect effects. The review might adopt the principles of the RIA approach, but look retrospectively at the regulations in operation and not just at proposals.

In Asia, the policy development of sustainability also follow such practice. Many governments conducted exercises on establishing the policy framework for sustainable development. The objective was to find out the options for the effective policy framework in order to implement sustainable development. The framework developed from these studies looked into the issues from three perspectives to arrive at the policy options: namely the policy priorities, policy instruments, and institution arrangements.

1 Policy priorities, which are a clear statement of the objectives, that is, what it is that needs to be achieved. Expressing objectives as "priorities" is important since not all the desired achievements are likely to be practicable within the foreseeable time-frame;
2 Policy instruments, which are used to achieve the priority policies;
3 Institutional arrangements, which are required to successfully identify and formulate the priorities for the policy, specify the appropriate policy instruments and implement them, and in due course, undertake their view (Figure 5.1).

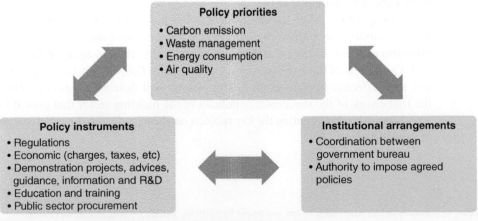

Figure 5.1   Policy Framework for Sustainable Development. © Arup

This framework provides a tool to study the policies of sustainable development systematically, focusing on their rationale, implementation strategy, and effectiveness for all the various stakeholders. As many Asian countries share a significant similarity in urban context and some have similar legal systems, they could provide valuable information for the policy makers when they conduct the RIA of their own.

5.3   POLICY PRIORITIES

In recent years, policy proposals were discussed and pursued in some Asian countries without the existence of an overarching sustainability policy framework. To address the growing policy challenges, environmental problems, and conflicting interests, all available knowledge, as well as supporting techniques and procedures are helpful to promote the need for formulating policy-making.[68] The issue for governments, however, is how to ensure that the various approaches are coordinated properly and implemented consistently with the objectives of sustainable development. The developments of policy in Asian cities have shown some of the same characteristics and do share some similarities.

5.3.1   The moving target

Archiving sustainable development is a moving target. The key policy issues change with our understanding of the problem. The approach adopted by the European Union (EU) on formulating the priorities can be a good point of reference. Among other things it involves prioritising a limited number of sustainable construction actions.[69] Of course, EU priorities may well differ from Asian countries, although they are certainly relevant to local circumstances. The EU established a high-level committee to review the possible areas for prioritising actions and making appropriate decisions. To some extent, the Kyoto Protocol and other instances of international collaboration have fulfilled the role of information provider for the problems facing the building and construction industry in Asia. Regional and local forums were organised to discuss the policy issues of sustainable development and coordination of action.[70,71,72] There is no shortage of information on the issues to be dealt with. Top items on the list are climate change resilience, carbon emission reduction, environmental degradation, energy efficiency, and renewable energy, etc. The task is to preliminarily prioritise them before taking the necessary action (Table 5.1).

Issues concerning the political environment also affect the priority of policies of a nation. For some Asian democracies, the continuity of policy may face problems with a change of administration. Technologies and market competition can also affect priorities. For instance, the commercialisation of photovoltaic (PV) in the market and most recently internet of things (IoT) technology are changing the landscape of sustainable development in Asia.

Table 5.1 National Plans and Policy Priorities of Sustainable Development in Asian Countries

Issues of sustainable development	China	Japan	Singapore	Hong Kong
Climate Change	Law on the Desert Prevention and Transformation (2001)	Act on Promotion of Global Warming Countermeasures (1998);		Resilient (2015)
Carbon Emission	Energy conservation Law (2007)	Energy Conservation Law (2008)	Energy Conservation Act (2013)	Fuel-mix (2014); Energy efficiency (2015)
Environmental	Environmental Protection Law (1989)	Basic Law of Environmental Pollution Control (1967); Basic Environmental Law (1993)	Environmental Protection and Management Act (1999); Environmental Pollution Control Act (1999)	
Liveability	Revised Development Standards for Eco-County, Eco-City, Eco-Province (2008)	Japan's Strategy for a Sustainable Society (2007)	Singapore Green Plan (1992); Sustainable Singapore Blueprint (2015)	Sustainable Building Design (2011); Air Ventilation (2006)
Smart City	Renewable Energy Law (2005)	Renewable Portfolio Standard (2003)		
Others	Clean Product Promotion Law (2012); Government Procurement Law (2003)	Waste Management and public Cleansing Law (1970); Green Purchasing Law (2000)	National Recycling Programme (2001)	

5.3.2 Prioritisation of policy issues

Implementation of sustainable development is a complicated process. In particular, many cities have been developed in an environmentally exploitative manner for a long time. There are a great many problems that are equally pressing. Prioritising the limited resources to address these urban challenges is of vital importance. In parallel, globalisation and regional integration have also raised new challenges and opportunities that the regional economies need to consider when formulating their national policy on sustainable development so as to allow for better social, economic, and environmental development.

5.3.2.1 *Drivers of changes*

Globally, the policy objective of sustainable development is driven by the vision of a clean and healthy environment and a strong economy. In recent years, the policy priorities of the sustainability of some developed Asian countries have extending from purely environmental issues such as air pollution, carbon emissions, and energy use to issues of social sustainability such as liveability, mobility, and the connectivity of people. The key objective is to create communities that are more people-friendly as well as environmentally friendly. This concept has expanded from low-carbon development to sustainable development by encompassing the social and economic considerations, in addition to the environmental concerns. In the process of this green transformation, many other regional mega-trends have also driven the changes, including rapid urbanisation, economic integration, rising income, and the changing pattern of consumption.[73] These drivers of policy change have characterised the prioritisation of policy for sustainable development in the region. Some characteristics were observed in the regional leaders concerning their objectives when advocating for sustainable development:

1 Regional integration – for regional leadership
2 A top down national plan – for enhancing national competitiveness
3 Green GDP – for addressing environmental pressures
4 Bottom up public engagement – for inclusive society and social equity
5 Green consumption – for capacity building in the green transformation

5.3.2.2 *Regional leadership*

Asia has started to be recognised for its leadership in global output, trade, and development, which is attributed to its regional economic integration. Economic integration is most evident in East Asia (Japan, South Korea, Taiwan, Hong Kong, China, and other ASEAN countries). Intra-regional trade, as a share of East Asia's total trade, accounted for 56% in 2006.[74]

The region now has the opportunity to take the lead in sustainable development with the benefit of this regional integration. Regional trade and investment frameworks and responses to common regional challenges (urbanisation, energy security, and resource scarcity) facilitate joint investments in strategic sustainable development. Emission trading systems, for instance, would deliver more environmental and economic benefits if the geographic coverage was larger. Policy changes that are perceived to affect a country's economic competitiveness in trade, such as carbon tax (or even carbon trading), will have a greater chance of succeeding if harmonised at the regional level. When the geographic scope is wider, changes are more effective and cost-efficient. Also, with globalised production systems and the global movement of capital, good practices in sustainable development can easily spread across the nations of the region, such as through trade and investment.

Being a regional economic and technology leader, Japan has much to share with other countries. In the post-World War II period, Japan has sought to establish their political and economic leadership across Southeast Asia. Its effort to maintain strong political and economic relations with countries in the region could help Japan promote regional cooperation for achieving sustainable development. It could also be beneficial for Japan in assuming regional leadership in encouraging sustainable development for greater regional prosperity (Example 5.1).

Example 5.1   A lesson learnt from the policy initiatives of Japan

Japan has the longest history of energy conservation in Asia. Driven by resource scarcity, Japan started implementing energy-saving measures as early as the end of the Second World War. Back then, economic growth had been the main target so as to catch up with the more advanced economies. In July 1949, the Industrial Standardisation Law was established. Based on this the Japanese Industrial Standard (JIS) Research Association was set up. During the first oil crisis in 1973, Japan relied on oil much more than any other developed country. In late 1973, Japan's Petroleum Supply and Demand Optimisation Law set oil-supply targets and restricted oil use. With low levels of natural energy resources and a subsequent dependence on imported energy, the two oil crises that occurred in the 1970s provided a key trigger for the introduction of the Law Concerning the Rational Use of Energy (Energy Conservation Law, Law No. 49) in 1979.[75] The government reorganised the policy system toward energy security and a nationwide effort on energy consumption, conservation, and efficiency was practiced. Over time new goals were added into the Energy Conservation Law and it saw revisions in 1993, 1998, 2002, 2005, and 2008.

After experiencing problems with water and air pollution during 1950s and 1960s, Japan began to strengthen its environmental policy through regulations over the environment. These were enforced under the Water Pollution Control Law and the Air Pollution Control Law, and resulted in a reduction of pollutants during the 1970s and 1980s. In the 1990s, as global warming became a very important environmental problem, Japan was active domestically in taking measures to mitigate the emission of greenhouse gases. Long-term low growth forced the Japanese to reconsider the real meaning of economic growth. They began to become concerned with living quality and to establish the relationship between energy conservation and environmental protection. As a result there is less resistance to promoting a low carbon economy.

Afterwards, the viewpoint of the global environmental protection was becoming coherent. The United Nations Conference on the Environment and Development (UNCED, the Earth Summit) was held in Brazil in 1992 and an agreement on climate change was passed. This in turn led to the Kyoto Protocol, which was passed at the COP3 (the third Conference of Parties of the UN Framework Convention on Climate Change) held in December 1997. It was here that each developed country agreed to reduction targets of greenhouse gas (GHG) emissions. The Kyoto Protocol was not only a milestone in the history of energy saving, environmental protection, and in the reduction of carbon emissions, but it also confirmed Japan's position in the world economy as a leader in the field.

5.3.2.3   Top-down national plan: Green to gold

A top-down approach is generally adopted by all countries when addressing climate change issues, with the government setting the national plan and a road-map for the priority and implementation of policies. In Asia, countries like China and Korea have incorporated targets for sustainable development into their long-term as well as interim five-year plans. One driver of change is the enhancement of economic competitiveness. The introduction of the Building Energy Efficiency Regulations in these countries is a good example. When the China's Building Energy Code was first implemented in 1995, it set a target of a 50% improvement compared with existing buildings. When it was revised in 2005 the standard was raised to 65% for implementation in major cities such as Shanghai and Beijing. Similarly, the use of non-fossil fuel was set at 5% in 2001, which will be raised to 20% by the year 2020. These polices have provided a stimulation to the market and generated demands for green products. Innovation in building design has blossomed with demand getting larger and larger.

Many Asian countries have treated this green movement as an opportunity to transform their industries to be more market competitive. This first happened in Japan, and was then followed in Korea and China. Other small economies have excelled at green building design, capitalising on their current advantages in the market. Japan and Korea are using this movement to create a regional leadership position through various collaboration efforts with other nations in the region. This is exemplified by the Green Growth Initiative of Korea.[76] In 2008, the Korean Government formulated a policy of "low-carbon, green growth" (LCGG) as the pillar of a new vision for the economy. Korea's green growth (GG) strategy was formulated initially as a 50 trillion Won Green New Deal, to help the country get over the 2009 global financial crisis. More signifi-cantly, it was framed in terms of a five-year green growth plan (5YGGP) which has since been vigorously promoted from 2009-2013, as the centrepiece of the government's growth and development strategy. The Korean government views climate change not as a cost but as an opportunity, and their strategies for dealing with it are couched as industrial policies, designed to stimulate the development of new green industries equipped with green technologies – conceived as new "growth engines" for the economy (Example 5.2).

Example 5.2   The green growth initiative in Korea

Green growth seeks sustainable growth by reducing greenhouse gas (GHG) emissions and environ-mental pollution. The principal objectives outlined covered those associated with a low-carbon society and energy security; new engines of industrial growth; and enhanced quality of life combined with international leadership.[77]

A Presidential Committee on Green Growth (PCGG) was subsequently formed in February 2009, as a high-level coordination committee, bringing together representatives from the principal ministries involved – including Finance, Industry, and Resources – as well as from the private sector and academia. The Korean National Assembly passed the Framework Act for Low-Carbon, Green

Growth at the end of 2009.[78] This has provided the legal framework for all the subsequent initiatives, including the following two focuses related to buildings:

1 Green cities, for example, green transport, green buildings, fast rail as well as green land management;
2 Promoting green consumption and lifestyle, for example, eco-labelling and carbon-certification.

The low-carbon green city project was initiated on President Lee's suggestion at the Gangwon Development Discussion in 2009. The Provincial Government (Gangwon-do) accepted the suggestion and formed a Green Growth Board. Finally, Gangneung City was selected for the low-carbon green city project.

The low-carbon green city project aims to develop low-carbon green cities to proactively cope with future climate change, and provide a carbon-free, green urban environment. The project also aims to set globalised models for future urban form and a green society. A conceptual model consists of the following six components or features: (1) Land Use & Spatial Structures, (2) Ecology & Green Systems, (3) Low-Carbon Energy & Housing, (4) Resource Cyclical System, (5) Green Transportation, and (6) Green Lifestyle.

One of the key initiatives is the Four Rivers Restoration (4RR) project on the Han River. It was built on the site of the previous impressive water engineering project known as Cheonggyecheon, which had to do with the recovery and restoration of a significant waterway through the heart of Seoul. It was championed by Lee Myung-bak, who was then Mayor of Seoul.

There has been much (ill)-informed criticism in the foreign press (as well as the Korean press) of Korea's GG strategy as amounting to little more than a water engineering project. This is because the Presidential Office and the PCGG have tended to put more emphasis in their public statements on the 4RR project. It is clear that there is a great deal of overlap between China's twelfth Five-Year Plan (2011–2015) and Korea's 5YGGP (2009–2013).

5.3.2.4   Green GDP

Environmental catastrophe is another driver for a change to sustainable development in Asian developing countries. China's economic success has come at the expense of air pollution, surface water and groundwater contamination, deforestation, resource depletion, growing waste, and simmering social unrest.[79] This environmental breakdown is partly due to policies which emphasise economic growth over other goals. In 2004, China decided to implement a "Green" GDP programme with the political shift in priority towards a more environmentally sustainable economic model. As buildings are the main energy consumer across all sectors, effort has been turned to the building of sustainable (or low carbon) developments in China's urban transformation. The idea of the 3Rs (Reduce, Reuse and Recycle) or a circular economy has been advocated in the planning of new towns. For China, low carbon urban transformation is even more critical as they are now the largest oil importing and carbon emission country in the world. Further uncontrolled urbanisation will only make the situation worse. With strong leadership and support from the Central Government, the policy of Green Urbanism in China is moving in the right direction, the principles of which are now being put into practice. Over 100 eco-cities are now being built in cities across China. Many of them are being developed in collaboration with other countries, including the Sino-Singapore eco-city in Tianjin, the Sino-Japan eco-city in Caofeidian Beijing, and the Sino-Germany eco-city in Tsingtao.[80] The UK is also working with China on developing the city-level framework for the implementation of eco-cities in four major cities. The model of executing high-level targets has now filtered down into the practices of the building industry, beginning with urban planning, and then followed by building design and construction. The efforts of the building industry over the past few years have amassed a great deal of valuable experience on tackling the challenges that they are facing. The policy of building more low carbon eco-cities has transformed the industry into following green practices and building the required capacity for wider adoption in the coming years in order to meet the need for further urbanisation in China, and across the region as whole (Example 5.3).

Example 5.3 China's policies on climate change

In September 2014, the Chinese Government published its action plan against climate change for the period 2014–2020. The action plan includes the following overarching goals, to be achieved by 2020: (1) to completely execute the action plan for carbon emission control, (2) to make substantial achievements in low-carbon pilot projects, (3) to strengthen adaptability of climate change, (4) to build skilled labour capacity, and (5) to strengthen international exchange and cooperation. Key goals that are focused on buildings are summarised as follows:

- Energy conservation – to control energy consumption of the steel and building materials industry, by 2020, the Authority will aim to manage carbon emission not to exceed the level set for the twelfth FYP; to accelerate the transformation of existing buildings to low energy consumption mode. In 2014, China's energy consumption for unit gross domestic product and carbon dioxide emissions decreased by 29.9% and 33.8% compared with that of 2005. The emission reduction targets set in the twelfth FYP have been successfully met. China has become the world's top country in using new energy and renewable energy. Between 2011 and 2015, energy intensity (energy consumption per unit of GDP) fell by 18.2% and carbon intensity declined by 20%.
- Low-carbon development – to strictly carry out the whole cycle of low-carbon construction, for example, to build energy-efficient gas, electric, heat and water facilities, and sewage and waste management facilities, to make regulations of the building's life year, to stipulate a building's life-cycle management, demolishment management, to utilize high-quality low-carbon steel and concrete materials to increase the building's life span, to promote green wall and green roof, to strengthen lighting management, and to implement the "users pay" policy. Figure E5.3.1 shows the Eco-city Proposals in the plan.
- Green building – to utilize geographic advantage to produce solar and geothermal energy, to exercise compulsory solar energy implementation at certain solar-energy intensive area, to promote green labelling and standards such as the LEED and Three Star certifications, in order to achieve the government's goal for green buildings to achieve a 50% share of all new buildings by 2020.

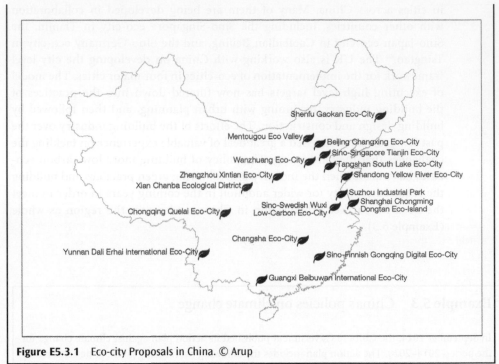

Shenfu Gaokan Eco-City

Mentougou Eco Valley

Beijing Changxing Eco-City

Sino-Singapore Tianjin Eco-City

Wanzhuang Eco-City

Tangshan South Lake Eco-City

Zhengzhou Xintian Eco-City

Shandong Yellow River Eco-City

Xian Chanba Ecological District

Suzhou Industrial Park

Shanghai Chongming

Chongqing Quelai Eco-City

Sino-Swedish Wuxi
Low-Carbon Eco-City

Dongtan Eco-Island

Yunnan Dali Erhai International Eco-City

Changsha Eco-City

Sino-Finnish Gongqing Digital Eco-City

Guangxi Beibuwan International Eco-City

Figure E5.3.1   Eco-city Proposals in China. © Arup

5.3.2.5   *Bottom-up approach: Building from the ground up*

The bottom-up approach is gaining more and more ground in developed countries and economies across Asia in arriving at the creation of sustainability policies. These policies, in particular, those related to the environment of the local community, have to receive support from the local society itself. The bottom-up approach has also been widely introduced as a counterbalance to the national culture, and primarily aims at empowering the local population to participate in whatever development process may have conflicts.[81] Otherwise the issues may cause political and legal disputes between the government and the public. The continuity of policy is particularly crucial for democracies in Asia, as policy should not be affected by a change of administration. Regular and in-depth consultation with stakeholders is an important part of the process in arriving at new policy initiatives. Singapore and Hong Kong are examples that favour a bottom-up approach for formulating the strategies for sustainable development. For any development, there will be conflicts between interested parties, and this will affect the decisions on distribution of resources such as land, and the sharing of responsibilities such as waste handling. Involving the community on project design and implementation can reduce resistance from the public. A balanced approach is critical for any policy formulation, in particular, in selecting the instruments to best use for the execution of policy. Enhancing the liveability of urban space is a policy issue for developing countries in the region (Example 5.4).

Example 5.4 Singapore's sustainable blueprint

Singapore intends to reduce its emission intensity by 36% from 2005 levels by 2030, with the aim to peak around 2030. The approach encompasses power generation, buildings, households, industry, and transportation sectors. A key step will be to progressively change the fuel mix to 10% fuel oil and 90% natural gas. Regulations govern large energy users, household appliances, vehicle carbon emissions, and building sustainability. Incentives include energy efficiency financing schemes aimed at encouraging companies and households to invest in energy-efficient technologies such as the Grant for Energy Efficient Technologies (GREET) or the Energy Efficiency Improvement Assistance Scheme (EASe).

Besides the traditional top-down approach, the Singapore Government has also engaged the community through consultations, exhibitions, surveys, and dialogue sessions. More than 130,000 people participated in the exhibitions and consultations, whereas almost 550 people gave their views. This was amalgamated in the Sustainable Singapore Blueprint that outlined plans for a lively and liveable Singapore. The latest 2015 Sustainable Singapore Blueprint[82] reported on the progress of achieving the targets for 2020 and 2030. It also outlined the national vision and plans for developing into a more liveable and sustainable country. Three key areas on the agenda for sustainable development were identified, namely the Liveable and Endearing Home, a Vibrant and Sustainable City, and an Active and Gracious Community.

With the Liveable and Endearing Home plan, the government is committed to put the effort into building more sustainable housing estates. Key measures are to be incorporated into the design of Housing and Development Board's (HDB) flats. These include a pneumatic waste conveyance system, solar panels, centralised chutes for recyclables, rainwater harvesting, rooftop greenery, and elevator energy regeneration systems. Some of these sustainable features will also be introduced to existing HDB estates.

With the Vibrant and Sustainable City plan, the government aims to optimise the urban planning of the city state. The approaches to be employed include integrated planning processes, decentralising jobs from the Central Business District to the regional commercial centres, and exploring the greater use of underground spaces. Specifically concerning waste management, Singapore will work towards becoming a Zero Waste Nation, with a target of a national recycling rate of 70% by 2030. To promote green buildings, a greater penetration of the Green Mark scheme is planned with a target of 80% of buildings to be Green Mark certified by 2030.

With the Active and Gracious Community plan, the government is doing a lot of work encouraging environmental education in schools, engaging the community through various green events, encouraging more green volunteers, and making better use of crowd-sourcing to draw innovative ideas from the community. Each district will develop their green targets and plans for their own districts.

5.3.2.6 Green consumption

Buying less, using less disposable items, choosing green products, lean construction, and supporting green business models are all known effective behaviours for businesses and individuals to achieve the objective of a low carbon economy. Yet, it is difficult to realise in practice, even in developed countries. Green consumption is becoming a key policy issue because its effect on reducing carbon emissions is immense. The market for green products is not yet well established because the market has not been provided with sufficient information. To rectify the problem, governments in many Asian countries are working on a labelling scheme for various products, including those in relation to energy consumption, water consumption, and low carbon construction materials. Many products to do with energy efficiency are regulated in a way to only meet the minimum standards required. If they fail to meet these standards, they are not allowed on the market. The Mandatory Energy Efficiency Labelling Scheme (MEELS) in Hong Kong and Singapore have marked over 25 housing appliances and pieces of office equipment. Consumers are now provided with the necessary information to arrive at an informed decision when selecting products. Other than these items, the public can also choose to buy or rent premises that have a green label. Buildings certified with a BEAM Plus (Hong Kong), CASBEE (Japan), Green Mark (Singapore), EEWH and CGBL (Taiwan), or a 3-Star (China) are now available on the market (Example 5.5).

Example 5.5   Hong Kong's climate change strategy

In September 2010, the Environment Bureau of the HKSAR Government launched a public consultation regarding "Hong Kong's Climate Change Strategy and Action Agenda", in which it proposed to adopt a voluntary carbon intensity reduction target of 50%–60% by 2020, as compared with the 2005 levels. The carbon intensity has further extended to 65%–70%, as compared with 2005 levels, by 2030 in the latest revision on January 2017. The strategy and the respective action agenda for mitigating greenhouse gas emissions include maximising energy efficiency to improving energy efficiency in buildings through a tightening of the Building Energy Code, and reducing the energy demand for air conditioning and other major electrical equipment. With the consultation period over the HKSAR Government is now considering the views and suggestions received from the public and is planning the way forward.

The consultation also proposed a voluntary carbon intensity reduction target of 50%–60% by 2020 as compared against the 2005 levels – this translates to an absolute annual emission level of 28–34 million metric tonnes of CO_2e (MMtCO_2e) in 2020, which translates to a 12–18 MMtCO_2e reduction from businesses as opposed to the usual increase (Figure E5.5.1).

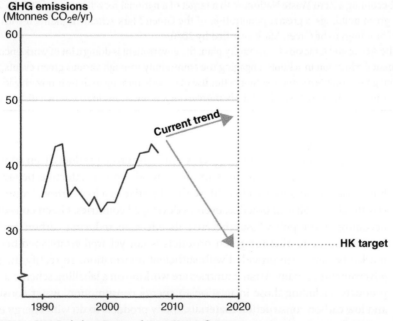

Figure E5.5.1   GHG Emissions Target of Hong Kong. © Arup

To achieve this ambitious target, appropriate action was suggested in the consultation document. The agenda spanned a wide range of sectors (power generation, buildings and construction, and transportation and waste). With the proposed action agenda, it was estimated that around 15 $MMtCO_2e$ (80% of the total reduction of 18 $MMtCO_2e$) of the emission reduction can be attributed to the change in fuel mix for renewable power generation, with the increase in nuclear power (increasing from 23% to 50% of the total electricity generated each year) accounting for the majority. This is in sharp contrast to the meagre 1 to 2 $MMtCO_2e$ (8%) in building (1.2 $MMtCO_2e$) and appliance upgrades (0.2 to 1 $MMtCO_2e$).

The action agenda proposed in the public consultation paper focused mainly on energy supply-side solution (i.e., the energy mix change), however, the development of nuclear energy is a controversial topic, and particularly so after the Fukushima incident. An important question to ask given the current climate is: "is it possible to adopt large scale renewable power generation?", or perhaps more appropriately: "what is the potential power generation capacity of Hong Kong?" Given the constraints in space available for renewable resources, more policy alternatives for reducing energy demand must be considered.

Since 60% of our greenhouse gas emissions in Hong Kong can be attributed to the operation of our buildings (25 $MMtCO_2e$), a large part of the reduction demand can be affected by making our buildings more efficient – buildings are the problem as well as the solution. Creating green consumption could be the best approach for resource-scarce Hong Kong.

5.3.3   The Asian way of change

It is difficult to generalise the approach adopted by Asian countries in prioritising the sustainable development agenda. Unlike Europe, there exists a huge disparity in the stage and intensity of urbanisation amongst the various countries in Asia. The social, economic, and environmental challenges faced, and the resources available to tackle the problems are always bound to be different. Different political systems as well as cultures draw them further apart when discussing and agreeing on solutions. However, the recent megatrend of regional integration has brought some hope for further collaboration. Although unlikely to work in a similar way to the EU,[83,84] the regional political blocks such as ASEAN and South-South can create the necessary platform for discussion on cross-border issues such as better resource management (air, water, and forests), adoption of green technologies, and capacity building or green practices. Collaboration can bring significant benefits to the region, developing and developed countries alike.[85]

5.4   POLICY INSTRUMENTS

Any policy will need an instrument in order to be implemented. There are several types of policy instruments that are applicable to achieve the objectives of a sustainable development policy. It is generally agreed that the government should be clearly in the lead when promoting policy. Other stakeholders should also be engaged in support of implementation. The choice of instruments is clearly an important element of the policy framework. In general, policy instruments can be

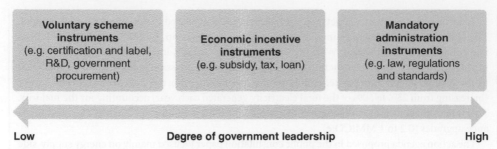

| Voluntary scheme instruments (e.g. certification and label, R&D, government procurement) | Economic incentive instruments (e.g. subsidy, tax, loan) | Mandatory administration instruments (e.g. law, regulations and standards) |

Low   **Degree of government leadership**   High

Figure 5.2   Policy Instruments and Government Involvement. © Arup

broadly classified into three categories, namely Mandatory Administration, Economic Incentives, and Voluntary Schemes (Figure 5.2).[86,87] They help to regulate operation and practices in the market as well as control the business activities and behaviour of individuals in order to achieve the required policy objective.

Regulations have considerable scope to influence behaviour, in particular, when the issues of control are new to the stakeholders. "Command and control" provides a direct mechanism for change in the targeted parties and is the most favourable option for Asian countries. Many new laws and regulations have been put in place in recent years. However, there is the drawback of high implementation costs. Governments are required to maintain resources in order to enforce the regulations. They also need to collect a lot of data to gauge the effectiveness of the measures. Regular reviews on standards and requirements are necessary, which has proved to be too expensive for developing countries. For example, the U.S. and Japanese energy standards are revised every few years, whereas the standards in China have not been revised since its inception in 1995.

Experience in the EU has also found that the most direct route to changes in behaviour can be achieved via influencing incentives, that is, by impacting the industry's profitability. Governments have at their disposal a range of instruments that are applicable in different contexts. However, economic measures and public procurement are likely to have the greatest impacts on behaviour as they directly affect industry "bottom lines" through costs and demand for construction products. Where possible, governments should consider the possible options for further introduction of economic instruments with the "polluter pays" principle in order to achieve sustainable construction and to prioritise the areas for application.

Other instruments will have more focused impacts on behaviour. Demonstration projects, advice, guidance, information, and Research & Development all have a place in the government's policy framework. This is an area where the government can give the private sector support. Public procurement as a policy instrument is also widely mentioned in international literature. However, there are some major obstacles to be overcome.

The following paragraphs discuss the main problems and provide some directions for the selection of instruments by the government in the pursuit of sustainable development policies.

5.4.1   Regulations and standards "The Stick"

The issues of sustainable development are both complex and multi-faceted. It requires an effective regulatory system to control the behaviour of all stakeholders by distributing the rights and responsibilities in the fairest way possible. In Asia, ordinances are the most commonly used instruments at the present time, having been introduced over many decades so as to regulate and control the industry. Regulations provide instructions and mandate the way forward or by controlling behaviour. For the building sector, the areas of control are far-reaching, covering the planning of development, building design and energy efficiency, and green operation (Figure 5.3).

Traditionally, the objectives of building regulations are to safeguard the health and safety of the inhabitants. Fire and structural safety are the main concerns, together with the basic life-safety provisions such as ventilation and daylight. In recent years, the energy efficiency of building design has been incorporated into the building control items in response to the global trend on carbon reduction. Building regulations in China, Hong Kong, Singapore, and Japan have incorporated into the requirements for building designs a reduction in the energy consumption of the building. Since 1995, when China produced its tenth Five-Year Plan, they have issued various national standards and codes for energy efficiency, such as the Civil Building Energy-Saving Design Standard (JGJ26-95) in 1995, the Residential Building Energy Saving-saving Design Standards in Hot summer and Cold Winter Zone in 2001, and the Residential Building Energy-saving Design Standards in Hot Summer and Warm Winter Zones in 2003. These standards prescribed the design for, among other things, building disposition and form, and window and wall insulation. Similarly, the building codes of Singapore and Hong Kong have incorporated the calculation of the Overall Thermal Transmission Value (OTTV) of buildings in order to regulate the energy loss due to operation of air-conditioning in non-domestic buildings. In recent years, many governments in the region have also started to focus on controlling the production and sale of products that consume energy

Figure 5.3   Regulations for Planning, Building Energy, and Energy Efficiency. © Arup

and water and that emit pollutants during operation by setting mandatory requirements for efficiency. This helps to promote green consumption in society.

5.4.2   Economic instruments "The Carrot and Stick"

As the market for green products and services matures, there are more and more economic instruments being developed by governments worldwide to implement green initiatives.[88,89] These are policy instruments that have the intended effect of influencing industry practices through impacting on costs or demand, and/or by providing incentives. Most obviously these include taxes, charges, subsidies, grants, and so on, which are financial in nature. But economic instruments could also include non-financial reward schemes such as fast-tracking the processing of selected applications, gross floor area (GFA) allowances, and so on. The role of the government in direct control of the operation of industry is minimal regardless of which type of economic instrument is adopted.

The intention behind economic instruments is to try to incorporate, as additional costs to designers, developers, and users of buildings and structures, a reflection of the full cost to society, and hence to attempt to influence behaviour in a way which is consistent with society's objectives. For example, landfill charges are currently being proposed to recover the full costs for disposal of construction and demolition waste. When introduced, these charges will impact the quantity of waste generated since waste-generating parties will have an incentive to minimise the impact of the charge placed on them. Moreover, charges will shift the costs of provision of landfill facilities from the general taxpayer to those directly responsible for the waste, thereby reducing the need for further taxation.

Another example is water. The price of water in Hong Kong does not reflect the full costs of the service. Not surprisingly there are complaints about the considerable wastage of water.[90] A full cost recovery policy for construction would undoubtedly reduce water utilisation and wastage, and conserve resources. The "polluter pays" principle is well established in the literature as well as in policy. This is the case not only in environmental literature, but also in worldwide sustainability practices. The principle could be further exploited for the good of a society in relation to construction. There are limited situations where charges can be introduced but they would significantly affect the achievement of sustainable construction since they would have a great deal of positive beneficial effects (Table 5.2).

5.4.3   Voluntary schemes instrument

Under voluntary schemes, the government has no control on the operation of the industry nor can it provide any incentives to facilitate business. It is a market solution that relies on the market force alone to be sustained. It normally occurs in mature economies where the flow of information is unobstructed and allows the market to regulate itself.

Table 5.2 Selected Economic Instruments for Sustainable Development in Asia

Country	Economic Instruments
Singapore	Innovation for Environmental Sustainability (IES) Fund (2001) Energy Efficiency Improvement Assistance Scheme (EASe) (2005) Clean Energy Research and Testbedding Program (CERT) (2007) Water Efficiency Fund (WEF) (2007) Clean Development Mechanism Documentation Grant (2008) Design for Efficiency Scheme (DfE) (2008) Energy Research Development Fund (ERDF) (2008) Grant for Energy Efficient Technologies (GREET) (2009) 3R Fund (2009) Sustainable Construction Capability Development Fund (SC Fund) (2010) Pilot Building Retrofit Energy Efficiency Financing (BREEF) Scheme (2011) One-Year Accelerated Depreciation Allowance for Energy Efficient Equipment and Technology (ADAS (2012)) Tax Incentives for Renewable Energy (2014) Green Mark Incentive Scheme for Existing Buildings (GMIS-EB) (2015) Green Mark Incentive Scheme—Design Prototype (GMIS-DP) (2015)
Japan	Japan Fund for Global Environment (JFGE) (1993) JPMorgan Japan Technology Fund (2005) Japan's Voluntary Emissions Trading Scheme (JVETS) (2005) Green Vehicle Purchasing Promotion Program (2009) Eco-Car Tax Breaks (2009) Feed-in-Tariff Scheme for Renewable Energy (2009) Funding Program for World-Leading Innovative R&D on Science and Technology (FIRST Program) (2009) Global Environment Research Fund (2010) Environment Technology Development Fund (2010) Grant-in-Aid for Scientific Research (about stabling a sound material-cycle society) (2011) Tax for Climate Change Mitigation (2012) Low Carbon City Act (Incentives for low carbon city planning and implementation, low carbon building – residential and non-residential) (2012) Green Fund (2013) The Act on the Improvement of Energy Consumption Performance of Buildings (Building Energy Efficiency Act): Incentive Measures (Voluntary): Exception of floor-area ratio regulation (2016)
Korea	Environmental Improvement Fund (1992) 21st Century Frontier R&D Program (1999) Eco-Technopia 21 Project (2001) Environmental Venture Fund (2001) Feed-in-Tariff Scheme for Renewable Energy (replaced by the RPS) (2006) Environmental Industry Promotion Fund (2009) Fiscal Incentives for Renewable Energy (2009) Recycling Industry Promoting Loan (2012) Green Building Incentive Ordinance Including fund, tax exemption etc. (2013) Renewable Energy Financing Programme (2015, 2016) Building energy saving design standards (2016) Seoul Renewable Energy Requirement (2016)

(Continued)

Table 5.2   (Continued)

Country	Economic Instruments
China	Technologies R&D Program (1986)
	National Hi-tech R&D Program (863 Program) (1986)
	Innovation Fund for Technology-based Firms (1986)
	Mobilizing financing from National New Products Program & National Key
	Technologies R&D Program (1986)
	National Basic Research Program of China (973 Program) (1997)
	China CDM Fund (2006)
	Renewable Energy Development Fund (2008)
	Green Carbon Fund (2008)
	The funding for energy efficiency renovations in North China (2008)
	Golden Sun Program (2009)
	Feed-in-Tariff Scheme for Renewable Energy (2011)
	Green building direct subsidies (2012)
	Tax Rebate for Wind Energy Producers (2013)
	State financed grant for green building demonstrated city (2013)

Participants are incentivised by recognition in the public and business communities. The recent success of green building labelling across the region is a good example of this measure. The market needs neither regulation nor incentives from the government (in most cases) to operate. Another example is the implementation of corporate social responsibility (CSR) in the business sector. This is a form of corporate self-regulation integrated into the business model. Listed companies on the Hang Seng Index in Hong Kong are now encouraged to disclose the actions taken for CSR in their operation.

5.5   INSTITUTIONAL ARRANGEMENTS

The construction industry is a complicated system that involves many players. For the government, how to channel the "command and order" of regulation effectively (with minimum effort) to these parties will decide if the regulation is successful or not. Conflicts will always arise when authority and responsibility is not well-defined. Similarly, for a market solution, effective flow of market information is crucial to prevent "market failure". Barriers are not uncommon and it is critical to set up an arrangement within government to talk through the policies and motivate the various parties.

5.5.1   Hierarchy – who to lead

On the top tier, should government or the private sector take the lead? Clearly, government must play a leading role in the development of policy through its over-arching interests in sustainability, but it is evident that the private sector is both interested in, and wishes to make progress with, encouraging sustainable development. Combining both sets of interests, both public and private, into a fully co-ordinated approach would be highly desirable as a way of taking green policy forward. This conclusion was arrived at on the grounds that only the government has the overall commanding interests regarding the environment, heritage, energy use, and other aspects of sustainability.

However, the private sector has undertaken several initiatives and demonstrated an important commitment to the concept. The recent success of the voluntary green building label which drives green building development in the market is a good example. Also, many private companies in the region have taken the initiative in integrating sustainability measures into management and operational practices without resorting to government support or intervention. It is reasonable to suppose that the private sector will continue to build on these initiatives and will wish to continue to play an active part in the development of the policy towards sustainable construction. Indeed, it is entirely possible that the private sector will assume a leading role in providing ideas and proposals which are consistent with the overall objectives of the policy.

Some semi-governmental organisations, non-government organisations (NGOs), and professional institutions can also advise and make recommendations to the government on strategic matters, major policies, and legislative proposals that may affect or are connected with the construction industry. If they respond positively then there would be the prospect of an "intermediate" option, one of joint leadership between the government and the private sector on identification of the issues to be tackled as a top priority, and for guidance on possible solutions.

While it is certain that policy will benefit from having the support of the private sector, the government will still need to foster this support. However, the support of the private sector cannot be automatically assumed in all cases as there are entrenched interests on many sustainability issues which will resist change. This is clearly the case in relation to landfill charging for construction and demolition (C&D) waste, charges that reflect the resource costs for providing water to the industry, and conservation of heritage, for example. The government can overcome resistance by taking the lead and, where necessary, imposing policy changes for the wider good of society.

The following options are now being adopted by Asian countries:

1 Government takes the lead – an option favoured by most countries on the grounds of overall commanding interest
2 Private sector takes the lead – not likely to realise all the desired results, although active participation by the private sector greatly facilitates the development of green policy
3 Intermediate option of joint public/private sector leadership roles – dependent on the attitudes and priorities of the semi-governmental body as the only feasible body to assume the leadership role

5.5.2   Government coordination and authority

The organisational issue for the government is how the administration should manage the development of sustainable policy given its importance yet very wide scope.

The existing structures for the development of policy and policy administration for the building sector, and its upstream and downstream activities, have built up over the years in response to specific needs. The government administration's lines of responsibility are essentially functional. This is perfectly satisfactory as an arrangement while policy needs are highly focused and specific, such as in relation to housing, transportation, heritage, and so on. But in such a structure, over-arching issues may be missed or inadequately dealt with, and lateral issues that require input from many areas of government may receive scant or contradictory treatment if responsibilities are weak or non-existent.

The public remarked very emphatically that the government structure has become cumbersome and sometimes difficult to understand in this sector, and that dealing with the government in relation to regulations, for example, requires several points of contact and is confusing for the public.

Organisational issues need to be dealt with primarily at the policy or bureaucratic level. Lower down at departmental and agency levels, implementation responsibilities are clearly defined because the departments and agencies have highly specific executive functions. This is common practice for some Asian cities such as Singapore and Hong Kong (Example 5.6).

Example 5.6   A model institutional arrangement for changes in Hong Kong

Leaving aside the Central Bureau (i.e., Chief Secretary for Administrations Office, Financial Services and Treasury), the bureaus that have more than a minor interest in the building and construction issues are illustrated in Figure E5.6.1. In the figure, the part above the line illustrates the responsibilities of sustainable development policy in the government. This includes the Sustainable Development Unit (SDU) of the administration section at the working level and, feeding into it, the council at overall Sustainable Development (SD) policy direction level. The prospective involvement of the Construction Industry Council in sustainable construction is also shown (this is discussed further below). Below the line are the various bureaus with an interest in the construction sector whose distance from the sustainable construction box and relative sizes gives an impression of their importance to sustainable construction policy. Closest to the sustainable construction box are the Environment Bureau and Development Bureau, with their direct construction responsibilities and to a lesser extent the Transport and Housing Bureau. Further away, in the sense of having less direct interest and involvement are the Home Affairs Bureau, the Education Bureau, and the Labour and Welfare Bureau. Further still are other bureaus whose interests in construction issues are more specific, such as the Commerce and Economic Development Bureau, Innovation and Technology Bureau, and the Food and Health Bureau.

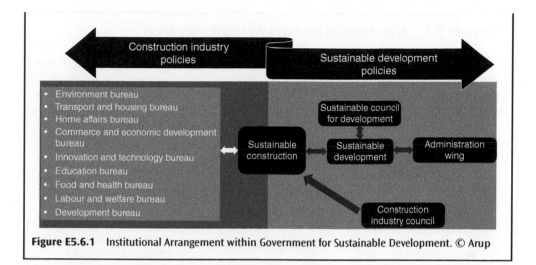

Figure E5.6.1 Institutional Arrangement within Government for Sustainable Development. © Arup

5.5.3 Proposal for eco-city implementation

The implementation of sustainable development in China is still at an early stage. Various policy instruments from regulatory approaches to economic incentives are in place. Yet the process is facing many problems. As stated by Qiu Baoxing, former vice minister of the Ministry of Housing and Urban-Rural Development (MOHURD), the fundamental problem is due to market failure. Delivering sustainable development to reduce carbon emission can be considered as property. The execution of property rights involves many owners, whose decision to conduct business is subject to many externalities in economic terms. For example, what is the cost and benefit that affects a party who did not choose to incur that specific cost or benefit? In the building sector, the implementation of sustainable development involves the developer as well as other external parties such as the users/buyers and the government (representing the interests of the public good). In this case, the public good is the environment and the well-being of the people. Under the current practice, the developer will normally enjoy the economic interests from building development, but the environmental pollution caused by energy consumption and waste generation from the buildings will not be borne by the developer but by the public, and the cost for cleaning up such pollution is usually paid by the government. Therefore, the Chinese government is taking the lead in rectifying the problems in the market and is guiding it on a more sustainable course. A new institutional arrangement based on the cost and benefit needs to be developed and implemented. Cost models must be put in place to evaluate the cost and benefit to all stakeholders and make it transparent to all parties (Example 5.7).

Example 5.7   A cost model for eco-city implementation

A recent study has demonstrated that the market failure was caused by a "monopoly market" and "information asymmetry". The core information on the true costs and savings of the project, both financially and environmentally, are not known. Therefore, a cost-saving analysis model for the project was established.

The model (Figure E5.7.1) is useful in formulating the institutional arrangements, in terms of the responsibilities and benefits to the different stakeholders of the development. The principle of redistribution of cost and benefit amongst all stakeholders is important.

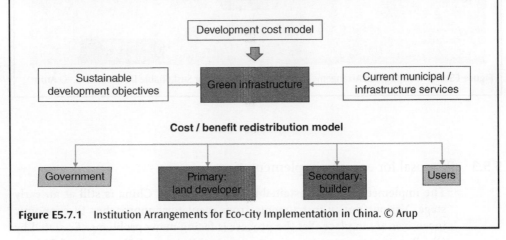

Figure E5.7.1   Institution Arrangements for Eco-city Implementation in China. © Arup

5.6   SUMMARY

Proper governance is pivotal to the success of implementing sustainable development. At the government level, a policy framework should be in place to direct the stakeholders. At the top of the framework, policy priority should be formulated. There are many drivers (external and internal) that influence priority in Asia. In developed countries, creating business or technology advantages is the priority, while developing countries will have to focus on addressing the environmental challenges created by rapid urbanisation. Policy instruments should also be in place to facilitate implementation. Other than regulations, economic instruments are becoming more popular to engage the key stakeholders.

Chapter 6
Policy implementation

6.1 INTRODUCTION

The issues surrounding sustainable development affect everyone in society. Implementation of any policy initiatives on these issues will involve many parties, and are not limited to simply the government. Some successful cases do not even have much involvement from the government. It is a complicated exercise that needs strong leadership from the government and appropriate instruments to execute the policy. Experience in Asian countries has found that regulation is more common as the policy instrument, as its effect can be swiftly realised across all parties. It does, however, have the drawback of high cost at implementation. Some governments have sought the help of businesses and allow the sharing of the leadership role with private sector companies and NGOs. The solutions from the market can either be incentivised by the government or the market itself. The recent development of green construction across the region is a good example. To sustain this green transformation, governments and business leaders have another role to play – to building more capacity for all parties who support the green movement. Education and training are key factors in sustaining the activities involved. This chapter will look into the processes of policy implementation and the successful cases that happened in Asian countries on addressing the problems of urbanisation that led to the adoption of sustainable development.

6.2 GENERAL APPROACH

The challenges in addressing the issues of sustainable developments, such as environmental problems, economic growth, poverty reduction, social development, equity, and living standards are complex and inter-connected. Governments are much better at institutionalising sustainability than any other sectors, and if we are looking to achieve large-scale transformation, the government sector is the

Building Sustainability in East Asia: Policy, Design, and People, First Edition. Vincent S Cheng and Jimmy C Tong.
© 2017 John Wiley & Sons Ltd. Published 2017 by John Wiley & Sons Ltd.

obvious one to select. The United Nations Development Programme (UNDP) observed that active and effective governance requires institutions which are capable of delivering reliable and quality service by directing the talent, creativity, and resources of people and business societies in shaping the changes needed.[91] If public funds are required, the government can allocate and invest wisely, as well as manage public goods, including land and other natural resources equitably, for the benefit of all. Therefore, an integrated policy and budget are critical to allocate resources to strategic national goals, rather than to individual ministries or departments.

Business clearly also has a critical role to play, and has demonstrated its capacity and desire to advance the sustainability agenda. The private sector, however, cannot make the transition from a waste-based (resulting from our consumption behaviour) economy to a low-carbon economy without the necessary rules to regulate and facilitate the market so that economic life does not destroy the planet that provides us with food, air, and water. The private sector can produce the goods and provide the services efficiently, and innovations as well as efficiency are what the private sector is good at. Cross-sector partnership can only happen with the involvement of the government in the public-private partnership model.

East Asia's dynamic economic performance has provided the opportunity for green transformation. Regional economies have developed their own models for implementation arrangements, and this has resulted in different extents of government and private involvement. The options for the implementation of policy initiatives for sustainable development are summarised below with the projected outcomes (Figure 6.1).

In practice, the government needs to consider if the policy issue requires regulation to limit/constrain the operation from the market/business or individuals. For issues that the market is not well-informed about or that are highly subject to external forces, the government needs to have regulations in place. Environmental issues, to which the industry will try to avoid their responsibility, are a good example. In recent years, Asian governments have also looked into the market forces.

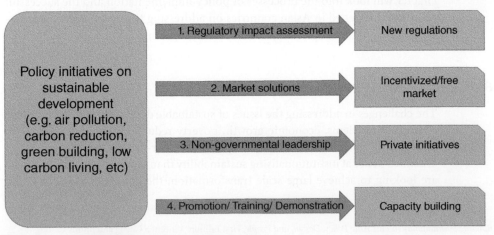

Figure 6.1   Options for Policy Implementation. © Arup

Green products are being promoted for wider penetration into the market place. An incentivised market for renewable energy applications and free market products such as green buildings are blossoming in Asia. Their success has also facilitated the formation of many NGOs and civil groups that help to take the lead role in society for championing the green movement. Such forms of private initiatives may receive better support from the public because NGOs have a better relationship with the public. An area that both the public and private sectors share the responsibility for is in building the capacity of green consumption. Professional institutions have been actively involved in the training of professionals on the "know-how" of delivering green projects. The green building profession is a new discipline that is now practiced widely in Asia.[92] Its existence provides the momentum to transform the market of green building products. With regard to enhancing the demand for green products, the education to the public is an ongoing work to increase the buy-in from the mass. No single measure is enough, they have to be implemented at the same time so as to achieve the policy objectives.

6.3   REVIEW OF THE REGULATIONS

Regulations are costly, at least in administrative terms.[93] Their introduction requires legislation, whether amending existing laws or formulating new ones. As in many Asian countries, some governments have put a great deal of effort into furnishing the regulations on containing environmental performance such as equality impact assessment (EQIA), indoor air quality (IAQ), waste management, and controlling the behaviour of individuals in energy use and waste generation. Governments are regulating business practices in an attempt to create a low carbon society.

With such a regulatory approach, the government is further obliged to enforce the rules and this in turn involves more expenditure. Mainly the costs for regulation are imposed on industry and society and not on the government. In recent years, the government has adopted so many new regulations that a Regulatory Impact Assessment (RIA) needs to be undertaken. Typically, a RIA consists of identification and a clear statement of purpose for the regulation, its risks, fairness, and benefits which, if possible, should be given a value. However, the main issue for a RIA is in assessing the costs imposed on society as well as any other indirect effects.

In practice, there is always a case for reviewing the regulations which affect buildings and their construction. The review might adopt the principles of the RIA approach, but look retrospectively at the regulations in operation and not just at the proposals. Such a review might be undertaken in two stages. Firstly, the review might look at the overall regulatory arrangements affecting the building and construction industry. This would provide the "big picture" to see whether and to what extent, the existing regulations were in terms of relevance, as well as their complexity and if there was any duplication. Secondly, selective RIAs could be undertaken for those building and construction regulations that appear most out dated or that are duplicated. Example 6.1 shows as illustration in Hong Kong.

Example 6.1   Review of building regulations concerning lighting and ventilation

In 1999, the Buildings Department of HKSAR conducted a review on the Lighting and Ventilation of the Building (Planning) Regulation Chapter 123. It took five years for the study team, led by the architect Anthony Ng, Arup and CUHK to complete the review. Performance-based requirements concerning lighting and ventilation performance were recommended for domestic buildings in Hong Kong in lieu of prescriptive requirements. A practices note (Figure E6.1.1) was subsequently prepared to facilitate the implementation of the regulations. [94]

As a result, the initiative has changed the practices of building design. Practitioners are provided with more flexibility to plan for the layout of buildings and are able to protect the future building occupants' right to receive sufficient ventilation and daylight, therefore increasing their well-being and health. To rationalise the requirements, the team conducted a great many surveys to identify the needs of ventilation and daylight in specifically in relation to the high-rise and high-density urban environment of Hong Kong.

Daylight calculation

1. Set cone angle up to 100° symmetrical & perpendicular to window plane.

2. Calculate UVA bound by obstructions.

3. Maximum cone length = height of façade.

4. Count UVA to the boundary of adjacent property (including road & permanent road/open space, etc.).

5. For a low building in front of the window with a vertical obstruction <30°, count the area on as UVA.

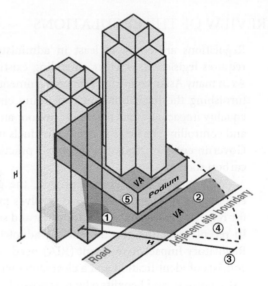

Figure E6.1.1   Performance-based Requirements for Daylight Regulation. © Arup

6.3.1   Building energy regulations

In the process a RIA might identify room for new regulations. The formation of the building energy efficiency regulation is a good example. In the past, the energy efficiency of a building was not an issue in Asia, partly because energy consumption in building sector was not significant as compared with other sectors such as industry and partly because the environmental issues at the time were not as pressing. Governments adopted voluntary measures to raise public awareness, and incentives in the form of rebates on electricity bills to encourage energy saving from building owners. Most of the buildings built in those days were poorly insulated and the systems were inefficient. Since the 1990s, however, governments have realised the importance of building energy efficiency in the reduction of national energy consumption and have followed in the footsteps of developed countries in relation to the regulation of energy use in buildings by controlling their design. Starting with China in 1995, building energy regulations have become prevalent and have proved to be an effective means in curbing energy consumption. They also form the key component in achieving the target of carbon reduction. The energy performance of equipment in buildings is also an important factor in many countries, as equipment consumes a lot of energy (Example 6.2).

Example 6.2 Development of Building Energy Policy in Hong Kong and Singapore

Formulation of the Building Energy Efficiency Ordinance in Hong Kong

Since 1998, the Hong Kong Government has been operating the Hong Kong Energy Efficiency Registration Scheme for Buildings to encourage voluntary compliance with the Building Energy Code (BEC). The BEC provides prescriptive minimum energy efficiency standards on lighting, air-conditioning, electrical, and lift and escalator installations. It is also a performance-based approach which allows innovative design of energy efficient installations in buildings as an alternative for compliance with the prescriptive requirements.

In order to further promote building energy efficiency, a new Buildings Energy Efficiency Ordinance (BEEO) for mandatory implementation of the BEC was passed by the Legislative Council in November 2010. The BEEO came into full operation in 2012. The core parts of the BEEO concerning the requirements for compliance with the BEC in the new construction of specified types of buildings and in major retrofitting works of relevant building services installations, as well as the requirement for conducting an energy audit for commercial buildings, are now within the grace period. This is to allow various stakeholders and the public to have ample time to adapt to, and become familiarised with, the requirements of the new legislation before it is put into full implementation.

The BEEO covers both public and private buildings, including commercial buildings, institutional buildings, community buildings, hospitals, passenger terminals, and the common areas of residential buildings and industrial buildings, and so on. New buildings of the specified type constructed after enactment of the BEEO will be required to obtain a Certificate of Compliance Registration (COCR) from the Electrical and Mechanical Services Department (EMSD) upon construction completion of the building, in that the concerned building services installations shall be certified for compliance with the BEC by a Registered Energy Assessor (REA). The COCR should be renewed every 10 years afterwards to ensure that the building services installations are maintained to relevant energy efficiency standards.

If major retrofitting works are carried out in a unit or common area of a prescribed building, the concerned owner or responsible person will also be required to obtain the Form of Compliance (FOC) from a REA upon completion of major retrofitting works to certify that the building service installations covered by the major retrofitting works are in compliance with the BEC. The BEEO also requires that an energy audit shall be carried out for central building services installations in commercial buildings once every ten years in accordance with the Energy Audit Code (EAC) issued by the EMSD.

It is estimated that the new legislation will result in an energy saving of 2.8 billion kWh, or reduce carbon dioxide emission by 1.96 million tonnes, in the first decade from those new buildings constructed after the new legislation has come into effect. Further energy savings will result from major retrofitting works in existing buildings as relevant building services installation will be required to comply with the energy efficiency standards as stipulated in the BEC, and the improvement works that may be implemented by building owners following the identification of various cost effective energy management opportunities through energy audits of commercial buildings.

Sustainable Building Policies in Singapore

In 2005, the first Green Building Masterplan was launched to push for new green buildings. When the second Masterplan was developed in 2009, a target was set for 80% of buildings to be green by 2030. By 2014, 25% of buildings were already green and the third Masterplan was developed with the vision for continued leadership, wider collaboration and engagement, and proven sustainability performance.

New and existing large buildings more than 2000 m^2 undergoing retro-fitting are legislated by the Building Control (Environmental Sustainability) Regulations 2008 to meet a minimum energy efficiency standard and instrumentation for cooling systems, on top of a minimum Green Mark score. The overall thermal performance of a building envelope to minimise solar heat gain is also regulated through the Code on Envelope Thermal Performance for Buildings. For projects with building cooling systems, the submission of air conditioning system information is required.

With the focus on existing buildings, the minimum environmental sustainability standard based on the Code on Environmental Sustainability Measures for Existing Buildings will be applicable to any single-use development comprising a hotel, retail building, or office building with a gross floor area (GFA) of 15,000 m^2 or more, when the building cooling systems is installed or replaced. A building cooling system will include any water-cooled or air-cooled chillers. Building owners are also required to submit yearly energy consumption to the Building Construction Authority, as well as perform a three-yearly system efficiency audit and compliance with design system efficiency.

6.3.2   Planning control for a better environment

Another control mechanism established in recent years is that of planning for development in urban areas. Like many other cities in the world, people in Hong Kong are increasingly concerned about the urban heat island phenomenon which brings additional discomfort to residents and pedestrians alike, particularly during the hot and humid summer months. Furthermore, with intense urban development over the past few decades, that often results in tall bulky buildings being in close proximity to each other, the heat energy stored in those buildings as well as heat generated by other sources at street level cannot be easily dissipated due to reduced wind penetration. To improve the urban living environment, the HKSAR Government has embarked on a number of studies to mitigate the less desirable effects of high-density living.

Firstly, in 2005 the air ventilation assessment system was introduced. This required that air ventilation effects arising from major government projects, planning for new development and comprehensive redevelopment, and the formulation or revision of major town plans must be studied. A set of design guidelines for the improvement of air ventilation in urban areas was incorporated into the Hong Kong Planning Standards and Guidelines. Though not compulsorily implemented for all private development, the scheme has been considered as the accepted practice for all planning applications.

Secondly the Urban Climatic Map and Standards for Wind Environment Feasibility Study was initiated in 2006 to review Hong Kong's urban climatic conditions to facilitate better planning and decision making at macro and district levels.[95] Technical investigation of the study revealed different urban climatic zones for different localities in Hong Kong and the need to plan and design our city at strategic and district levels to optimise urban climatic conditions. The study also suggested that it is feasible to develop a pedestrian wind environment yardstick. Together with the air ventilation assessment, these measures are now in place to improve the living environment at the planning level.

6.4   MARKET SOLUTIONS

One objective for implementing sustainable development is to reduce the use of energy and other resources as well as improve the quality of life in less developed countries. These targets can be achieved by enhancing efficiency or adopting various innovations. In principle, both strategies will result in savings that will be reflected in the life-cycle cost of a truly sustainable development. Such savings can be best realised by market actions that can create a new business for investment. In other words, a sustainable low carbon economy would be more attractive (and profitable) than a conventional waste-based economy. Yet, in practice there are some issues that cause failure in the market (refer to Chapter 5 for details). Government support is required to improve the business environment, in particular at the initial stage to improve the adoption and use of green products or services.

6.4.1   Incentivising the market

Governments have considered various possible options for further introducing economic instruments to achieve sustainable development and prioritise the areas for their application. Further research is required to establish the values of the society for environmental improvements, since it is necessary to undertake some measurements of these values when establishing the level of charges or taxes.

6.4.1.1   *Direct financial subsidies*

As discussed in Section 5.4.2, many economic instruments have been implemented in Asia to support green business. For example, both the governments of Japan and Singapore have offered financial subsidy incentives to Building Energy Efficiency (BEE) programmes. The Singapore Government has introduced "Green Mark Incentive Schemes" and since 2009, has offered 100 million Singapore dollars in assisting in the implementation of its Green Mark programme.

In China, the Ministry of Finance (MOF) and the Ministry of Housing and Urban-Rural Development (MOHURD) have launched two main financial subsidy programmes for retrofitting residential and commercial buildings, namely the "Interim Administrative Method for Special Fund for Government Office Building and Large-scale Building" and the "Interim Administrative Method for Incentive Funds for Heating Metering and Energy-efficiency Retrofit for Existing Residential Buildings". The National Development and Reform Commission (NDRC) and MOF introduced a subsidy fund programme in late 2007, targeting the promotion of high-efficiency lighting products. In 2009, MOF launched the first solar subsidy programme called the "Interim Management Methods on Financial Subsidy for Application of Building-Integrated Solar Photovoltaic". The government has also provided value added tax (VAT) deductions on energy-efficiency materials for walls and windows since 2001. These measures have proved to be very effective in boosting the market demand at the inception level of new business.

6.4.1.2   Gross floor area concessions

In Hong Kong, to give the green building movement more impetus in the private sector, in addition to the requirement for compliance with the Sustainable Building Design (SBD) Guidelines and the overall cap on gross floor area (GFA) concessions, all new buildings will be required to undergo the BEAM Plus assessment (but without mandating the rating obtained) if developers wish to apply for GFA concessions.

Development intensity in Hong Kong is controlled by limiting the permissible plot ratio and coverage of a building or buildings within a site based on its size and classification, and the height and use of the building. A plot ratio is a figure obtained by dividing the total GFA of the building by the site area. However, the floor areas of certain features such as green or amenity features, plant rooms, or car parks may be excluded from the GFA calculation. To contain the effect of building bulk while allowing flexibility in the design for incorporating desirable green or amenity features and non-mandatory or non-essential plant rooms and services, an overall cap of 10% will be imposed on the total amount of GFA concessions for these features. Certain features such as mandatory plant rooms, car parks, and communal sky gardens will not be subject to the overall cap on GFA concessions. The result of the BEAM Plus assessment as well as the estimated energy consumption of the common parts of a domestic development and of the entire non-domestic development will be published on the Buildings Department website upon completion of the development, to enhance transparency of information.

6.4.1.3   Environmental funds

In Hong Kong, the Environmental and Conservation Fund (ECF) established the Environmental and Conservation Fund Ordinance in 1994. Following the injections of $1 billion and $500 million into the ECF in early 2008 and mid-2011 respectively, the ECF has provided funding to support over 2,200 educational, research, and other projects and activities in relation to environmental and conservation matters. Under ECF funds, the Energy/Carbon Audit Project and Energy Efficiency Project aim to reduce building energy/carbon. The cash incentives will help to drive energy efficiency measures. The programme provides financial subsidies to private building owners to conduct energy/carbon audits and energy efficiency improvement projects in buildings.

Japan also has similar financial incentive programmes for promoting construction of green buildings, such as special fund for implementing CASBEE.

6.4.1.4   Public sector procurement

Governments should lead by example when it comes to the issue of green procurement. In Asia, many countries have a policy for the public sector to support the green market. For example, all new government buildings in Hong Kong are required to incorporate 2% renewable energy and the buildings themselves are required to achieve a "gold" level (the second highest) green building certification. This move is very effective in driving the market in the early stages of sustainable development by creating a "policy push" effect on the market. Once the market goes beyond the critical mass required for change, the market itself will provide the incentives.

Virtually all infrastructure provisions are procured by the public sector, and many Hong Kong buildings are also procured by the government or public agencies. Obviously the procurement procedures of the government can have important implications for the achievement of sustainable construction objectives. In April 2009, the HKSAR Government issued a comprehensive target-based green performance framework (the Framework) for new and existing government buildings, which set targets in various environmental areas, such as energy efficiency, renewable energy, indoor air quality, and greenhouse gas emissions. The adoption of comprehensive targets concerning the various aspects of environmental performance for government buildings has had a demonstrative effect in enhancing the environmental awareness of Hong Kong citizens and helps to fulfil Hong Kong's international commitment to reducing its energy use intensity.

Public housing procurement is another area that the government can influence the practices of the industry. People-oriented design can also be achieved if the public is involved in the process of policy formulation. Public housing in Hong Kong is a good example. Being the most densely populated city in the world, the need for an improved housing environment has never been greater. At the policy level, how to deliver sustainable housing in the most resource-efficient manner is the top priority for the administration. Other than deliver the prescribed numbers of housing units to meet the demand, the government also has the mandate to provide healthy and comfortable built environments for residents. Understanding the needs of the residents is critical for the government to optimise the design and provisions of the housing facilities. The practice of government public engagement is important in modern society to maintain harmony.

6.5   MARKET-BASED APPROACH

Since the early days of the modern environmental movement, there have been efforts to utilise economic principles to protect environmental goals. A free-market suggests that the state's function is to create and protect markets and systems of property rights, such that well-design markets will optimally allocate environmental resources so as to maximise human welfare.[96] Some successful cases have been found in Asian countries.

6.5.1   Green building certification

In recent years, green building certification has been one of the most successful voluntary schemes across the globe. It capitalises on market forces and resources from the private sector to drive the development of buildings towards being more sustainable. It has become a worldwide trend that the development of the green building movement has been championed by NGOs under the umbrella of the World Green Building Council. With the implementation of green building certification schemes or standards, these institutions prove to be effective in pushing the boundaries of the practice of sustainable building design. It has also created a market for green building construction in the industry and articulated effectively the complex issues of green construction to the public.

The development of green buildings can also be driven by the market itself, capturing its momentum and creating standards for practice. Green building labels have proven to be an effective tool used in the market place. The benefits of this market-driven instrument are manifold. Competition in the market can create new business for the survival of the eco-system of green buildings. Some leading green building label schemes in this region are highlighted in Chapter 7.

6.5.2   Sustainability report and index

The business community has taken the practice of sustainability seriously over past two decades. Corporations across all business sectors have been eager to produce sustainability reports that demonstrate their commitment to sustainable development to both internal and external stakeholders. In Singapore, of the 537 listed companies studied, 30% (160 companies) were reported that covered the topic on sustainability. In Hong Kong, 57% of the listed companies published an independent corporate social responsibility (CSR) Report. A sustainability report discloses information to the public on the corporation's economic, environmental, social, and governance performance which improves its transparency and accountability. To standardise the practices, guidelines called the Global Reporting Initiative enable all organisations worldwide to assess their sustainability performance.

A listed company is also encouraged to make use of a CSR report to attract potential investors. To facilitate operation in the market, a sustainability index was established. Launched in 1999, the Dow Jones Sustainability World Index was the first global sustainability benchmark. The index serves as a benchmark for investors who want to integrate sustainability into their portfolios, and it also provides an effective engagement platform for companies who want to adopt sustainable best practices.[97] In view of sustainability investment growing across the globe, the Hang Seng Corporate Sustainability Index Series was launched in 2014, which include Hong Kong-listed companies that perform well with respect to corporate sustainability. This has demonstrated the effectiveness of market forces to drive the green movement.

6.6   PUBLIC-PRIVATE PARTNERSHIP (PPP)

The general argument over whether the private sector should be involved in the provision of utility services to the public is wide-ranging and essentially political. It is possible to summarise some of the principles behind the desire, expressed by many governments throughout the world, to involve the private sector in providing public services, in particular on green projects.

First, it is often argued that the private sector is much more easily able to achieve efficiency improvements, a key factor for green development of which is that savings are difficult to realise, and that it is difficult to take full advantage of technological progress. In essence, this argument suggests that investors have more effective means of ensuring that the managers of a company act to reduce costs and increase revenues.

Second, on a practical level, the government could be under a number of constraints that would make it difficult to implement some infrastructure projects without private-sector involvement. Expenditure budgets might not be able to accommodate the significant investment needs of the projects. Locating and approving additional funds can take time, as can effecting any necessary administrative or institutional changes that may be required to facilitate running of the projects by government.

Involving the private sector in the delivery of what traditionally are seen as public services is a worldwide and ongoing trend. It is often argued that investors have more effective means of ensuring that the managers of a company act to reduce costs and increase revenues. It also recognises that the private sector has valuable experience in developing new products and services, managing risk, and interacting with capital markets. Experience of World Economic Forum (WEF) initiatives are particular valuable (see Chapter 4 for details).

Public Private Partnership (PPP) is an approach that public and private sectors share the responsibility for the delivery of services. In practice, the involvement of the government is mainly in defining the requirements of the service and providing the necessary institutional and regulatory supports to secure this service, whilst the private sector is responsible for delivering the service and capitalising on its innovation and flexibility. In this regard, risks can be allocated to the party best able to manage them. A typical PPP arrangement will last for a long period, between 10 to 30 years. The project will be managed by the private sector taking advantage of private sector management skills incentivised by having private finance at risk.

A PPP approach is thus higher up the list of private sector involvement than simple outsourcing or management contracts, or where government wholly finances projects up-front but it stops short of full divestiture.

Many governments throughout the world are willing to involve the private sector in providing public services. Government typically use PPPs for the followings reasons:

- Service delivery is more cost-effective by gaining access to experience, management skills, and management flexibility that may not otherwise exist in the public sector.
- Second, at a practical level, governments might be under a number of constraints that could make it difficult to implement some infrastructure projects without private-sector involvement. Expenditure budgets might not be able to accommodate the significant investment requirements of the projects. Locating and approving additional funds can take time, as can effecting any necessary administrative or institutional changes that may be required to facilitate running of projects by the government.
- Public service infrastructure finance be achieved with "off balance sheet" funding and, in budgetary terms spread the cost over its economic life.

For detailed advantages of PPP please refer to Example 6.3 about the typical key features of a PPP approach and the potential benefits.

Example 6.3   District cooling system implementation

Implementation of new technology requires support from the private sector. The experiences from the local and international practice of a District Cooling System (DCS) for both regulated and non-regulated markets have created some successful models on the procurement and operation of DCS. In specific, these models can be categorised in the following types:

- Base mode - the government (or its agent) will be responsible for funding, construction, and operation of the DCS. This is a conventional procurement method for public works.
- Operation and Maintenance (O&M) Agreement - the government will be responsible for the funding and construction of the facilities, whereas the DCS will be operated by the contracted parties. The tariff of the DCS service will be set by the government.
- Operator model (energy services contract) - the government will be responsible for the funding and construction of the facilities, whereas the operator will be responsible for the operation of the plant. Tariffs are set by the government and the operator can charge to the building users.
- DBO (Design, Build, and Operate) - the DBO contractor is responsible for the construction and operation of the system with the funding provided by the government. Tariffs can set by the government or negotiated between the government and the contractor.
- Concession/BOT (Build, Operate, and Transfer)/BOO (Build, Own, and Operate) - the BOT contractor will provide the finances, construction, and operation of the DCS. Tariffs are negotiated between the users and the BOT contractor under unregulated business or have to comply with the regulations if in place.

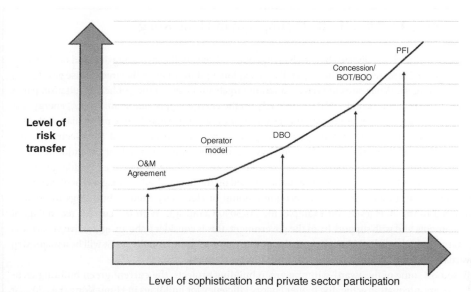

Figure E6.3.1   Level of Risk Transfer for Different Procurement Arrangements. © Arup

As shown in Figure E6.3.1, different procurement arrangements will create different levels of risk for the contractor/operator of the DCS and will end up with more and more sophistication on the control of operations by means of contractual arrangements.

From the perspective of the owner of the DCS, application of these models will depend on availability of funding and the risks that the owner is prepared to take. Project specific issues such as the design, programme of development, and uptake of potential users will also be considered.

6.7   COLLABORATION WITH PRIVATE SECTOR

Issues of sustainable development are multi-sector and multi-party. Involvement of the private sector in driving the development of sustainability is not a new phenomenon. In the 1970s, environmental groups were very active worldwide in influencing the policies regarding environmental issues. This resulted in many changes to the policies.

The issues of sustainable development are also multi-faceted and they require the engagement of the public for discussion. Support from civic societies is particularly important when implementing policy and there have been instances when a bottom-up approach was adopted when faced with controversial issues. Some governments partner with the groups concerned and treat them as a think tank. They could in turn initiate new policies and champion the movement itself.

The development of certifications and labels in the market has led to the establishment of various agencies to promote green building construction. The Hong Kong Green Building Council (HKGBC) was established and comprises representatives from the construction industry as well as other professionals. Their objective is to help raise public awareness of green building construction and to facilitate exchange and technological co-operation between Hong Kong

---

Example 6.4   The HK3030 Campaign in Hong Kong

To cope with the global climate change challenge and to deal with the increasing demand of electricity consumption, Hong Kong has the imperative to explore options to meet the greenhouse gas (GHG) reduction targets. While current GHG reduction targets proposed in the public consultation paper by the HKSAR government in 2010 mainly rely on the energy supply side solution (i.e., energy mix change), the development of nuclear energy has become a controversial topic particularly after the Fukushima incident. Since buildings in Hong Kong consume over 90% of the electricity, and account for over 60% of the GHG emitted citywide, there is a vast potential for the building sector in contributing to the GHG reduction targets. Therefore, the Hong Kong Green Building Council (HKGBC) wants to propose a holistic approach based on the demand side management of electricity consumption to reduce 30% of the absolute amount of electricity used in buildings in 2030 as compared to the 2005 levels, for example, the HK3030 Campaign. With the current technological developments and with support from the government, this is consider to be an aggressive yet achievable target, which will lead Hong Kong to be a more responsible metropolis and it will be a major step towards a sustainable built environment for future generations.

Based on some of the award-winning green building projects, the current green building technologies are adequate in achieving higher energy efficiency for buildings in Hong Kong (i.e., 50% or more efficiency when compared to the 2005 level). With the projection of increasing building stock and higher energy consumption per capita, the HK3030 Campaign is equivalent to a reduction of 52% electricity consumption when compared to the Business-As-Usual (BAU) consumption.

---

and the rest of the world. The HKGBC promotes green building standards and engages the community, industry, and government to create practical solutions for Hong Kong's built environment. It also creates a competitive international market for the green construction business. It is an over-arching body that includes all stakeholders and organisations involved in the various facets of the green buildings. Similarly, Singapore has established the Singapore Green Building Council (SGBC) (Example 6.4).

6.8   CAPACITY BUILDING

A great deal of attention has been drawn to the capacity building needs when addressing sustainable development.[98] The effectiveness of capacity building is enhanced through building abilities, relationships, and values that will enable organisations, groups, and individuals to improve their performance and achieve their development objectives. With regard to practicing sustainability, the role of government, as well as professional bodies, is the key when trying to influence the public and train people to behave more responsibly.

6.8.1   Demonstration projects and research and development

This broad category of for a policy instrument is relatively cheap to sponsor but at the same time are unlikely to have any dramatic effect on industry behaviour. As far as research and development (R&D) is concerned, foreign experiences can only be used up to a certain point. While U.S., UK, European, and other technical research into the construction industry may be generally applicable to Asia, it is felt that the unique circumstances for construction in the region requires specially directed research to ensure that technical solutions are appropriate.

The development of zero carbon building (ZCB) is a good example. The technologies for achieving ZCB are available in the market and many Asian governments have built their own ZCB to evaluate the applicability of those technologies to local conditions. It was found that climate is a determining factor for adoption of ZCB. Hot and humid conditions in Asia have resulted in the poor performance of many older existing technologies. Efforts have been put into more R&D work to adapt the technologies and promote further innovations. Demonstration projects of ZCB are found in Japan, Singapore, China, Hong Kong, Taiwan, and Korea.

6.8.2   Education and training of green practitioners

The study into the practice of green construction in the region has found that the lack of sufficient and competent professionals practicing green in the market is one of the key reasons for the slow adoption of sustainability in society.[99] The accreditation and training of professionals for green practice are pivotal to the success of a wider adopt of the green movement in society. As a result, some governments in the region have taken proactive action to equip their professionals with new knowledge and best practice.

The Singapore Green Mark Professional - To facilitate the implementation of the Green Mark scheme in Singapore, the Building Control Authority (BCA) has established the accreditation of Green Mark Professionals, namely the Certified Green Mark Manager (GMM), Green Mark Professional (GMP), Green Mark Facilities Manager (GMFM) and Green Mark Facilities Professional (GMFP). The purpose of establishing this specialist scheme is to recognise professionals who have attained a good foundation of knowledge. They can be employed for projects and provide their professional knowledge in the design and operation of environmentally friendly buildings. As an incentive, projects that a principal member who is a Certified GMM/GMP/GMFM/GMFP takes part in can earn Green Mark points.

The Hong Kong BEAM Professional - In light of the Practice Note issued by the HKSAR Government for making BEAM Plus the only official assessment scheme for new building development to obtain GFA concessions, the Hong Kong Green Building Council (HKGBC) has run the BEAM Professional (BEAM Pro) training scheme since April 2010 to help facilitate the implementation of the BEAM Plus project. BEAM Pros are green building professionals trained and certified by the HKGBC in all aspects of the entire green building life cycle. To become the accredited BEAM Pro, applicants have to undergo training and pass an examination. A key role of BEAM Pros is to inject the latest green building standards and practices into everyday building planning, design, construction, and operation. In a typical example, a BEAM Pro will be embedded into the project team. A BEAM Pro will be responsible for advising the team on how to achieve the required credit points to attain the client's desired green building rating level.

The Hong Kong Registered Energy Assessor (REA) - The Buildings Energy Efficiency (Registered Energy Assessors) Regulation formulated under the Buildings Energy Efficiency Ordinance (BEEO) came into operation in March 2011. Applications for registration as REAs were immediately opened up. The role of a REA is to assist the developers, owners, or responsible persons (such as tenants or occupants) of the prescribed buildings to comply with the statutory requirements of the BEEO. REAs are required to (1) certify the building services installations comply with the Building Energy Code for the developers, owners, or responsible persons; and (2) conduct energy audits for owners of commercial buildings. As all certification and auditing works under the BEEO are required to be carried out by REAs, a new profession in the field of energy efficiency engineering emerged when the BEEO came into full operation in September 2012. Through energy audits of commercial buildings, opportunities for the improvement of energy efficiency by retro-fitting and/or upgrading existing building services installations for energy saving can also be found.

The Japanese CASBEE Accredited Professional – The CASBEE Accredited Professional Registration System was established in 2005. There are three licences for CASBEE Accredited Professionals, (1) Building Accredited Professionals, (2) Detached House Accredited Professionals, and (3) Accredited Professionals for Market Promotion. To become a CASBEE Accredited Professional, applicants need to take a training course, pass the examination, and complete the registration process. Candidates for the Building Accredited Professional certification must be a Japanese first-class architect in order to take the examination. There were over 12,000 CASBEE Accredited Professionals as of March 2015.[100]

6.9   SUMMARY

Regarding policy implementation, the application of the right instruments by government is the key to success. At the early stage of green development, regulations should be in place to control business operation and the individuals involved. When a green market has formed, the government can then consider applying economic instruments, which has been proven to be effective in incentivising the market. Some businesses, such as green construction has now started working effectively as a market solution.

The institutional arrangements are also of equal importance. Normally, a government-led situation is common and there should be the right mechanisms to link up the policy creation process with its implementation. The private sector can also take the lead, and in most cases support from civic groups is also important.

Section 3
On design

Chapter 7
Sustainability transformation

7.1   INTRODUCTION

Putting the change framework discussed in previous chapters into practice is the next challenge for realising sustainable development. Transforming the practice of the building industry is particularly important. Sustainability as a discipline for practice is still at its infancy in the building sector. In some Asian cities, many experiences have amassed owing to a vast amount of construction activities occurring throughout the region over recent years. Some good practices are becoming mainstream, which has helped to escalate the penetration of green buildings in the market. Yet, the green transformation process requires two key steps: engaging stakeholders and empowering practitioners.

To engage stakeholders, the life-cycle consideration of building design, construction, and operation processes are helpful in engaging multi-disciplinary professionals to work together. Sustainable buildings are now being designed with due consideration towards construction method and operation, while it is still on the drawing board. The expertise of builders and operators are capitalised in order to achieve an effective design.

To empower practitioners, an innovative approach is required to deliver sustainable building designs. Integrated design is key to realising sustainable concepts and targets effectively. Globally, practitioners have endeavoured to develop a universal system for design that can help minimise the negative impacts from buildings. Practices on issues such as energy, water, environmental pollution, and so on, have been standardised through international collaborations. Yet, building design is an interactive process that requires deliberation to fit into the context of its surroundings. Rather than formulating rigid green design manuals, many professional institutions have established guidelines for practicing building sustainability. These guidelines provide a reference for

Building Sustainability in East Asia: Policy, Design, and People, First Edition. Vincent S Cheng and Jimmy C Tong.
© 2017 John Wiley & Sons Ltd. Published 2017 by John Wiley & Sons Ltd.

architects and engineers to develop ideas and meet the constraints (mainly cost and time) of projects.

In parallel, the adoption of green buildings worldwide is fuelled by the promotion of green building rating systems in the market. Certified by LEED and other rating schemes are considered as a "must-have" for some high-end office developments. This so-called "green wave" has also affected the design of many buildings in Asia.

Standardising the practice of green building through the process of integrated design and green certification in the market is evolving in the region. The trend of globalisation and regional integration will further help to define the important factors in resolving the key issues of climate change and the problems resulting from the urbanisation.

7.2   GREEN TRANSFORMATION OF BUILDING INDUSTRY

In general, buildings consume 40% of energy and emit up to 60% of carbon emission in a city. A green transformation of the buildings industry is key to the success of implementing sustainable development. All countries in the region are actively working on policies and implementation arrangements to enhance the contribution on carbon reduction from buildings. To promote green buildings in the country, it is required to build a capacity of related stakeholders for their buy in and make it a standardised practice in the industry to capitalise on the market force.

Amongst the critical actions for capacity building, engaging stakeholders for collaborations and empowering practitioners for implementing the building life-cycle processes are the most important factors. The building sector is a complex system that involves many decision makers. To make this exercise effective, the right strategy has to be in place.

7.2.1   Engaging stakeholders

Green building involves many different stakeholders, including architects, engineers, contractors, construction managers, owners, building occupants, building operators, and government agents, all of whom must fully commit to a project to ensure the success of an integrated design and construction process, which is the bedrock of green building.[101]

The multi-disciplinary nature of the building industry needs a strong collaboration amongst the inputs from professionals. This intensive collaboration should begin at a project's inception and be maintained until the completion of a project, and even during the operation stage when interaction with building users begins.

At a project's initiation stage, developers/investors must be committed to green buildings. Studying cost-benefits can help make an informed decision on adopting green building designs. Planners are also required to work closely with authorities if any incentives are to be granted for a project. In many countries, this is a critical juncture for the initiation of green building.

During the implementation phase, a whole building approach also requires a solid coordination between designers, contractors, and operators on the critical issues of building. During the course of a building's life cycle, consultants, contractors, and operators are three critical parties as they design, construct, and operate buildings. A design team is used to make all the decisions related to a building's design. That being said, construction and operation teams have accumulated relative knowledge and should be more involved in the decision-making process for design. This has proven to be an effective approach of incorporating suitable systems and technologies. Collaboration between the three parties has helped to change the behaviour in the building industry.

For the operational stage, it is crucial to have a correct understanding of the interactions between occupants and a building's technology. This will allow an improvement of a building's operations and design. If occupants have different behavioural tendencies it may require alternative technical solutions. Thus, alternative technical solutions may in turn affect or change the behaviour of occupants in a building. Therefore, parties involved (consultants, contractors, suppliers, and operators) must cooperate to provide more suitable technical solutions for end-users in order to influence their behaviour.

In fact, the behaviour of occupants in buildings is a major factor, which influences operations and energy consumption. This influencing factor is considered as deterministic in design rather than stochastic in nature. In the chain of a building's cycle, the behaviour of all parties, from early designs to the occupants' behaviour during its operational stage, is of equal importance in determining energy use. For instance, whether to apply new technology in the design, whether to improve the commissioning protocol for design verification, and attitudes towards preventive maintenance all play a big part (Example 7.1).

Example 7.1   Organisation for a sustainable building project

A building project combines multi-disciplines together with the client to form an executive best solution, especially to achieve sustainable development. A team enables options to be developed sensibly and clearly. This arrangement also allows team members to focus on sustainability and potential integration challenges. Figure E7.1.1 shows how a multi-disciplinary team is formed and a team's input during a building project. Typically, a team consists of a development team, architecture, structure, MEP (Mechanical, Electrical, and Public Heath), specialists (facade, lighting, acoustics, transport, geotechnical, etc.), delivery experts (construction), and facility management (operation). With a team working together during the early stages and throughout different stages of a project, they can work together to define clear and quantified targets, goals and tasks, identify challenges, implement programmes, costs, schedules, and evaluate the impact on certification.

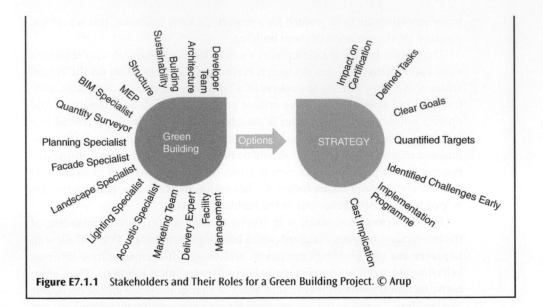

Figure E7.1.1   Stakeholders and Their Roles for a Green Building Project. © Arup

7.2.2   Empowering the practitioners

Lacking expertise in green building design, construction, and operation within the building professional circle is a key obstacle for green transformation in Asia's building industry.[102] In particular, a major factor for success related to green building is the use of an integrated process that involves an intricate, multidisciplinary collaboration between a wide range of stakeholders, all of whom work closely together to generate innovative and effective green building solutions. In recent years, Integrated Design Process (IDP) at the design phase has been a key for success in implementing green building strategies and technologies in a green building project.[103] A similar approach, known as an Integrated Construction Process (ICP), is also being practiced in the construction industry.[104] They provide a step-by-step approach to guide all practitioners to act accordingly to an agreed procedure at various stages of building development.

Practicing sustainability requires a more rigid process to standardise the operation. First, design standards have to be implemented for practitioners to follow consistently. Many green buildings built in the past contained features that were marketed as "green," but in fact were gimmicks or claims of environmental friendliness, or greenwashing. The industry should also be educated by good examples. In recent years, many successful green buildings have been built globally. Further collaboration within the region could help to develop a genuine model of green building for the region's climate, culture, and practices. Green practice should be enabled with appropriate technology. The region does not lack in products and skills in the market. The industry lacks in the integration of these technologies for more innovation, and Asian countries can learn from the developed western countries. Finally, practitioners should also be provided with feedback on applying green design. Stakeholder consultation will help in understanding the expectations of all key players. Recently, post occupancy evaluation (POE) has helped to gather feedback efficiency. Green building practice is still evolving, and further collaboration and cooperation can help to shorten the process for standardisation.

7.3   PRACTICE OF BUILDING SUSTAINABILITY

Modern architecture has been practiced for many decades. Yet, the concept of sustainable building is still new to many practitioners. A deteriorating built environment further complicates the practice of building sustainability. This is a result of rapid urbanisation in many regional countries. The industry needs to exercise a standardisation of practice in this area.

7.3.1   Definition of sustainable building

Multi-discipline and multi-party involvement throughout a life-cycle development is important to establish a framework to allow people and various processes to work smoothly. There are many objectives that have been developed worldwide on the planning and design of sustainable building development and its neighbourhood. Developing a fully comprehensive assessment system, which incorporates all of the aspects of sustainability, namely society, environmental, and economic sustainability, requires challenging a number of current practices and concepts.

During the compilation of the framework it became very clear that the comprehensive sustainability assessment tool must look wider than just building and the immediate site that the building is on. Urban planning is not strongly regulated in some countries in Asia and developers do not have to consider wider issues. For example, provisions of transport or public facilities including crèches, shops, and medical centres are requirements in many countries throughout the world. While this is accepted, the framework has been developed to include many of these social and economic criteria to encourage awareness of these issues and assist pressure for change.

While many design-based criteria are readily understood by designers and are quantitative, many of the social and economic criteria cannot be measured and will always be qualitative. For example, energy consumption can be measured, but this is not the case for the importance of cultural heritage (Example 7.2).

Example 7.2   Comprehensive environmental performance assessment scheme, Hong Kong

An exercise of developing a building sustainability assessment scheme was conducted in Hong Kong in 2005. During the development of Comprehensive Environmental Performance Assessment Scheme (CEPAS) a number of issues became clear. For example, a holistic assessment framework must not only consider the building itself but also its surroundings in which it is situated, that is, a neighbourhood. Many of the issues have been raised by considering a wider context and are not directly under the control of the developer or owner, which may not be adequately covered by legislation. One of the aims of CEPAS is to improve the weaknesses in the current system and help to promote change to deliver a greater degree of sustainability awareness throughout both the industry and the community.

These objectives together with the assessment issues or indicators (Table E7.2.1) define the framework or approach for CEPAS to assess sustainability at a construction level. The list below is organised around social, economic and environmental perspectives proposed for CEPAS.

Table E7.2.1   Objectives and Issues under CEPAS Sustainability Framework

Sustainability issue	Objective	Issues/indicators
Environmental	To enhance environmental quality	Reduce environmental impacts due to noise, pollutants, waste, etc.
		Create robust, adaptable an high-quality buildings
	To promote biodiversity	Enhance habitat diversity within the neighbourhood
		Realise urban "Brownfield" potential thus reducing countryside loss
	To safeguard natural resource	Manage energy and water resource sustainably
		Reduce demand for non-renewable resources
		Close local resource loops (re-use and recycling)
	To cut greenhouse gas emissions	Increase energy-efficiency of buildings
		Promote renewable
		Reduce car reliance and the need to travel
Social	To enhance local community	Create opportunities for local social groups and networks
		Strengthen social and cultural life
		Create an attractive public realm
		Promote local distinctiveness and value local heritage
	To increase equity	Promote accessible local facilities
		Enhance movement options – especially walking, cycling and public transport
	To promote safety and security	Encourage a sense of ownership and belonging
	To increase local decision-making	Building local social capital through the participatory process
		Create local partnerships and trusts
Economic	To promote wealth creation	Recycle financial resource locally
		Promote urban regeneration and renewal
	To promote employment	Enhance further education and training opportunities
		Ensure good public transport, walking and cycling connections to the wider area
	To reduce cost	Enhance efficiencies on resource use by means of designs and technologies

The indicators are developed based on internationally recognised indicators of sustainability (UNEP indicators, WHO Healthy City, and Urban Policy and SUDEV 21) with an adjustment for the local context of building assessment. The approach to develop CEPAS from these objectives and the sustainability indicator was to identify all of the performance categories related to a building and its surroundings, under these objectives and indicators. These categories were developed from a combination of experiences with developing building design and assessment tools internationally, and cross checking the issues identified using the SPeAR® tool.

The consolidated assessment structure focuses not only on the physical or environmental considerations, but also human-oriented considerations or necessities that defines a good living environment, as shown in Figure E7.2.1.

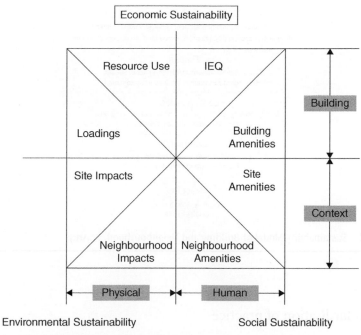

Figure E7.2.1   Sustainability Framework for CEPAS. © Arup

Eight performance categories were identified for CEPAS (see Figure E7.2.1). These were arranged separately to group building categories from those categories related to the neighbourhood context and also to separate physical categories from societal categories. This diagram more readily reflects building orientation issues rather than a more generic approach. As sustainability is very much a cross cutting issue affecting every aspect of a building's development, CEPAS has sustainability as a fundamental consideration, and each category of the diagram considers the impact of sustainability. Consequently, while the term "sustainable design" is not mentioned in the diagram, sustainability is considered within every criteria and sub-criteria in the diagram. The issues included to address social sustainability at the building level are intended to enhance the quality of services and amenities provided within the building/development to support occupants and the community. The "quality of life" is enhanced through an improved liveability. Those related to economic sustainability will cover the maintenance of the assets over a building's/development's life. A number of social, economic, environmental and natural resources indicators are included within the framework, which together cover all sustainability issues (Figure E7.2.2).

Without challenging softer issues, the change required in building standards and quality of life may not be achieved. There are a number of set challenges, particularly in the context of Hong Kong. The approach of CEPAS for the problem of applying judgment is by providing performance level guidance based on international best practice, modified by the views of local experts to suit Hong Kong's environment. Any qualitative marking system is open to challenge and the elements of CEPAS is no exception. The guidance provided is based on the independent views of numerous experts to provide a fair assessment system for the first version of CEPAS. As a number of assessments are carried out in Hong Kong, the guidance for assessing qualitative criteria may be reviewed in light of local experience. However, this should only be executed by independent experts to ensure that standards of assessment are not eroded by interested parties.

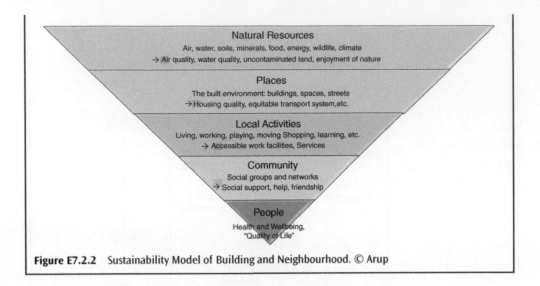

Figure E7.2.2   Sustainability Model of Building and Neighbourhood. © Arup

7.3.2   Standardisation of practice

Since a multi-disciplinary integrated design process requires closer interaction among parties for sustainable development, standard practice must be improved to adjust for new arrangements. Green standards and guides can provide practitioners with a good basis for best practice and provide non-experts with a general awareness of essential aspects.

7.3.2.1   *Green standards and guides*

Standards provide details on the requirements of what professionals need to design, from product material to the performance of buildings. ISO defines a standard as: "a document, established by consensus, approved by a recognized body that provides for common and repeated use as rules, guidelines, or characteristics for activities or their results." The requirements found in standards may either be prescriptive (identifying methods of achievements) or performance based (stating expectations of end results). Many international standards have been incorporated into the statutory requirements of the building control authority. These standards can be applied on products, design requirements, or management plans of related issues of green building practices. The widely adopted standards related to green buildings are summarised in Table 7.1.

In many countries, professional institutions are established to oversee the practices of professionals and provide support on the continuous development of various professions and its professionals. For example, AIA in the US has published a Guide for Sustainable Projects and RIBA in the UK has issued a Guide to Sustainability in Practice.

Table 7.1   Key International Standards Related to Green Buildings

Standards	Issue of concern	Country
ISO 14000	Environmental Management	International
ISO 50001	Energy Management	International
ISO 14040	Life-cycle Assessment	International
ASHRAE 90.1	Energy Standard for Buildings Except Low-Rise Residential Buildings	International
ASHRAE 62.1	Ventilation for Acceptable Indoor Air Quality	International
ASHARE 189.1	Design of High-Performance Green Buildings	International
BEC 2015	Building Energy Code	Hong Kong
SS 577:2012	Water Efficiency Management Systems	Singapore

7.3.2.2   Rating and labelling schemes

Green building rating or certification systems do not only focus on building products or equipment. The rating systems are building certifications, which provide acknowledgement or recognition to the achievements of specific environmental compliance or performance with a corresponding environment or building type.

Most of the leading rating schemes developed in Asia (as shown in Table 7.2) are quite similar in terms of assessment criteria. They address global environmental issues such as carbon emissions, and local issues such as water and energy efficiency. The use of materials along with the health and well being of occupants are also of concern. Example 7.3 has more details on CASBEE from Japan.

7.4   SUSTAINABLE BUILDING IN ACTION

The practice of sustainable building is a continuous process throughout the life cycle of a building, and requires all stakeholders to work in a collaborative manner.

7.4.1   Life-cycle consideration

According to the Whole Building Design Guide,[105] the main objectives of sustainable design, "…are to reduce, or completely avoid, depletion of critical resources like energy, water, and raw materials; prevent environmental degradation caused by facilities and infrastructure throughout their life cycle; and create built environments that are liveable, comfortable, safe, and productive." In practice, a building's life cycle can be divided into three stages, namely design, construction and operation. There exists many savings on carbon missions (or costs) if sustainable practices can be adopted (Figure 7.1). At a design stage, integrated designs are key to maximising the savings by considering various constraints and opportunities of the site and a project's nature. Designers are required to make informed decisions with the support of professionals and computer simulations. Usually, this is the stage when potential savings are at their highest. During the construction stage, material

Table 7.2   Available Local Green Building Rating in East Asia

Building Rating System	Type of Standard or Certification (First/Current Version)	Issues/Areas of Focus
Beam Plus (Hong Kong)	New Building Version 1.2 (1996/2012)	Site Aspects (SA) Material Aspects (MA) Energy Use (EU) Water Use (WU) Indoor Environmental Quality (IEQ) Innovations and Additions (IA)
CASBEE (Japan)	CASBEE-NC (New Construction) (2003/2014)	Indoor Environment Quality of Service Outdoor Environment (On-Site) Energy Resources & Material Off-site Environment
Green Mark (Singapore)	New Non-Residential buildings (Version 4.1, 15 Jan 2013)	Energy efficiency Water efficiency Environmental protection Indoor environmental quality, and Other green features
EEWH (Taiwan)	EEWH-BC (Basic Version, (2011/2015)	Ecology Energy saving Waste reduction Health Innovation
CGBL (China)	Assessment standard for green building GB/T 50378-2014 (2006/2014)	Land Saving and Outdoor Environment Energy Saving and Energy Utilisation Water Saving and Water Resource Utilisation Material Saving and Material Resource Utilisation Indoor Environment Quality Construction Management Operation Management Promotion and Innovation

---

Example 7.3   CASBEE, Japan

In terms of presenting the performance, the CASBEE of Japan has introduced a concept of quotient mechanism called Building Environmental Efficiency (BEE) by comparing both benefits (as referred as "Environmental Quality") and any adverse impact (as referred as "Environmental Load"), hence advocating "overall net benefit" brought forward by developments (Figure E7.3.1).

Environmental quality (Q) evaluates the enhancement of amenities for users in a designated area, where the outdoor environmental load (L) evaluates the negative aspects of environmental impact beyond a designated area. Should a project achieve net zero impact by scoring 1.0 or overall net benefit by scoring higher than 1.0, a project is given B+ (Good), A (Very Good) or S (Excellent), depending on its overall score. However, should a project achieve less than 1.0, meaning the project brings a negative net impact overall, it will be given a B- (Fairly Poor) or C (Poor).

CASBEE uses a complex weighting system to compute the score for the three sub-categories of Q and L. To determine the total score, the Building Environmental Efficiency method for urban development (BEEUD) is used. BEEUD divides the total environmental quality by the total environmental loadings.

Another unique attribute of CASBEE is the use of a three-tier criteria system, namely "Category", "Medium Category" and "Minor Category", with dedicated weighting factors for each tier. This illustrates the complex three-tier structure system of CASBEE Urban Development.

CASBEE is one of the most sophisticated schemes in the market (Figure E7.3.2). It has products that cover all scales of development, from a city to a single building, or an event and its materials. Some cities have their own version that incorporates the unique contexts of their building practices.

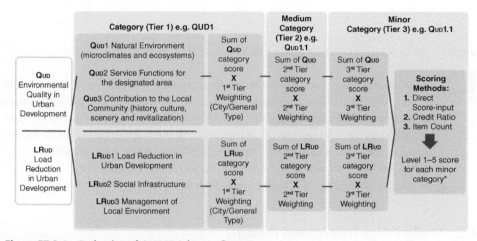

Figure E7.3.1   Evaluation of CASBEE Scheme. © Arup

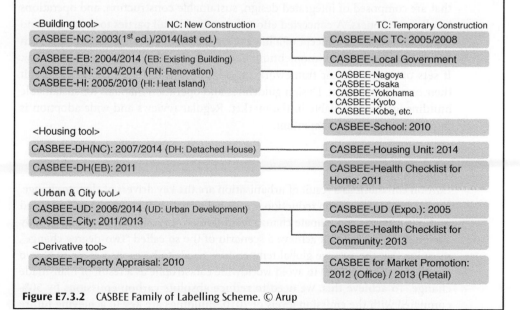

Figure E7.3.2   CASBEE Family of Labelling Scheme. © Arup

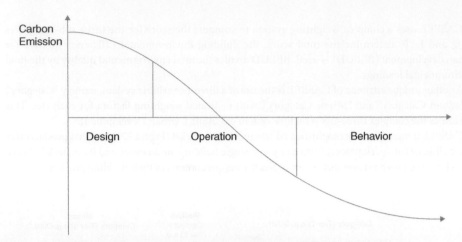

Figure 7.1   Life-Cycle Saving of Carbon Emission Opportunities. © Arup

use is key. Nowadays, sustainable materials are readily available in the market. With the help of LCA analysis, construction teams can work closely with project teams to achieve the lowest solution for carbon emissions based on material use. At an operational stage, the implementation of behaviour changes opportunities to the facility manager and the occupants are the key. Without proper management arrangements, buildings designed to be "green" may end up emitting more carbon than conventional buildings due to the unsustainable behaviour of occupants.

7.4.2   Design stage – integrated design

The successful realisation of sustainable building relies on an adoption of solutions that are composed of integrated design, sustainable construction, and operations by all practitioners. A concerted effort is required by all parties to set targets and meet deadlines. The concept of a life-cycle approach or whole building approach has gained the attention (momentum) worldwide for other practitioners to follow. It sets out principles or framework for adoption in different countries to suit their own environment. Design guidelines and operation manuals for sustainable building are now available in the market. Regular reviews and wide adoption is required to maintain momentum.

7.4.2.1   *Design target*

Carbon emissions as a result of urbanisation are the key drivers of climate change. Setting targets on carbon reduction is a global initiative. The IPCC has developed multiple scenarios on climate change for different reduction levels. It is generally agreed that we need that achieve a scenario of the so called "two-degree change", meaning that the average global temperature needs to be capped at a rise of two degrees by the year 2050 to avoid worldwide catastrophe as a result of man-made change. To achieve this, we need to reduce absolute carbon emissions by 50% compared with the emission levels of 2010. Buildings are key to achieving future

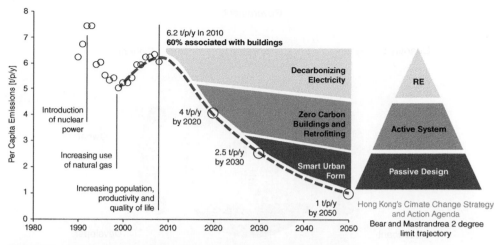

Figure 7.2   Interventions to Achieve 2°C Temperature Rise Limit. © Arup

sustainable development. Up to 40% of the world's energy each year is used to heat, cool, and light buildings. An energy efficiency reduction in the construction and operation of buildings offers the single most significant opportunity to reduce the impact of climate change. Figure 7.2 shows a possible strategy.

Understanding how far to practically push a design requires a thorough technical and cost assessment. Through this exercise, a designer has the ability to consider the following quantified features:

- How much energy, carbon, water or material reductions will a certain initiative deliver?
- How much will this cost and what is the payback/return on investment?
- How does this compare to other alternatives?
- How much does one need to spend to achieve a certain goal?

7.4.2.2   Closed-loop design process and collaboration

An integrated design approach is the first step to achieving the right solution to close the loop. Figure 7.3 presents a flow diagram, which highlights the design process required to enable a detailed energy/cost assessment. It should be noted that in order to make this process worthwhile, the concept design stage needs to be more detailed than is typical, involving close collaboration with all professionals including a project quantity surveyor.

Buildings provide safety, shelter, and comfort, and generate huge social and economic benefits. Yet, they also require land and use limited resources in the extraction of materials.

Designers share the responsibility with others working in construction to reduce energy and carbon emissions in the construction environment. This could mean designing new buildings to be as energy-efficient as possible, or applying sustainability criteria to the renovation of existing buildings. Greenhouse gas emission reductions, energy use, waste, construction materials, recycling, water

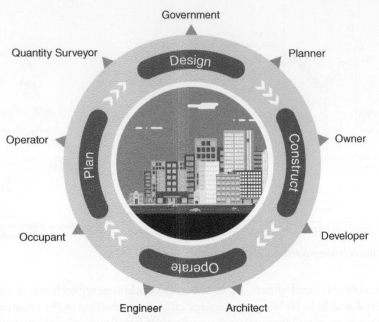

Figure 7.3   Closed-Loop Design Process. © Arup

usage, and the integration of buildings with infrastructure and social systems are all areas where responsible design can pay dividends.

If new and existing buildings were subjected to more ambitious standards of sustainability, up to 20% of current energy consumption could be saved. Yet, the sheer diversity of buildings, their different functions, uses, and intended life cycles make it a challenge to devise energy conservation measures for general use. A very different approach is required for a small family house compared to large residential developments, commercial offices or factories. The world's temperature has risen significantly in the last century, and 2005-2015 (just the last decade) was the hottest decade since records began. The International Energy Agency estimates that the world will need 60% more energy in 2030 than it does currently. If, as predicted, most of that demand is met by burning more fossil fuels, according to the best available climate science, it will create a very high probability of accelerating global warming. Caught in a vicious circle, our demand for energy is likely both to cause an increase in global warming and a continuing increase as a result. When the UK experienced record-breaking temperatures in July 2006, for example, the leap in demand to power air-conditioning units strained the national grid. The way to avoid this type of spiralling demand is to design buildings that balance comfortable indoor environment and reduced CO_2 emissions. Developing effective passive heating and cooling systems, which harness natural ventilation and shading to provide an alternative to power-hungry active cooling systems is one way forward.

It is possible to achieve comfortable summer temperatures in moderate climates using passive cooling measures, combining efficient building materials and modern design methods. Many East Asian countries fall into this climate zone. Increased solar shading, controllable natural ventilation and high thermal mass significantly decreases overheating, with a minimal extra increase in energy usage and carbon emissions. Such buildings may be difficult to realise in some regions. However, this is due to site restrictions, costs, external air quality, noise pollution or other constraints, particularly in the refurbishment of existing buildings and in urban areas. The work from the authors in adapted offices illustrates the success of a "mixed mode" approach, in which mechanical systems are used only when passive cooling systems have ceased to be effective – perhaps when ambient temperatures have risen to match interior temperatures. Proactive building management is also required to ensure that systems are used in the most effective way.

The precise changes ahead are unclear, yet the likely effects and implications for building designs are anticipated by means of adjusted half-hourly interior temperature profiles. These help to project the likely environmental and weather related demands on new buildings in the future, taking into account possible climate change scenarios. The best efforts to produce buildings along the lines of sustainable design stages will only work if owners, occupiers, and the government embrace change. Developments marketed as a sustainable lifestyle choice command as a premium have started in Asia and Australia. Yet, the greenest buildings can be designed are only as sustainable as its occupants will allow. Therefore, it is important to note that encouraging the behavioural change of occupants is key.

The challenge is for governments to promote and incentivise a better approach to buildings; for architects and building designers to continue to see sustainable development as a priority; and for the entire planet to make greener lifestyle choices. The buildings under whose roofs we live, work, and play need to be as sustainable as possible at every stage of their lifetime – from construction to demolition. This long-term view is no longer a luxury, but an imperative.

7.4.3   Construction stage – sustainable materials

For developed countries, buildings consume most of the material and energy consumption of a city, which also absorbs large economic resources and capital from investments. Measures that increase the sustainability of buildings must be encouraged and applied to various stages of building. Life-cycle assessment (LCA) is a concept to bring the horizon of looking at products and materials from cradle to grave, to categorise problems and assign priorities in finding a solution in order to minimise impact to communities and the earth as a whole, as shown in Figure 7.4, and to achieve the overall sustainability of buildings. LCA, as a tool, makes it possible for objective measures for all stakeholders of the building and construction industry, including the government, developers, architects, engineers, constructors and building operators, and so on, to assess the environmental impacts associated with the construction and operation of building and set targets for achieving the sustainable construction for the industry. The principle of LCA is to quantify the amount of life-cycle environmental impacts associated with all the processes of producing building materials, including extraction of raw materials, manufacturing processes, distribution, use and disposal. An LCA approach differs from other environmental management approaches in that it measures and calculates impacts normalised per unit of output.

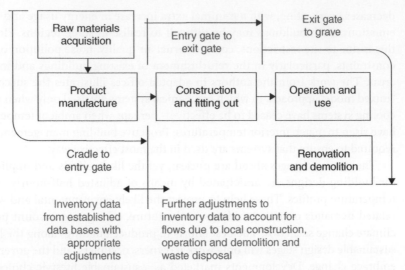

Figure 7.4   Stages of Building Life Cycle. © Arup

7.4.3.1   *LCA application*

LCA application currently remains limited in the region because a credible tool for the entire local industry is still lacking.[106] The essential features required by the LCA model, as seen in Figure 7.4, a building is regarded as a product and the product life cycle is divided into the following phases:

1   Cradle to entry gate - the cradle to entry gate phase starts from extraction of raw materials and embraces all the processes for producing the required construction materials, components, and logistics to the construction site. The inventory data required includes all the economic flows such as raw materials, intermediate products, recycled and reused materials, and the energy and ancillary inputs for the production and transportation of the construction materials and components. All the impacts (emissions or releases) incurred in these processes, which are the environmental interventions, are another essential part of the inventory data.

2   Entry gate to exit gate - the entry gate to exit gate phase corresponds to the construction phase of a building. This phase will consume materials and generate construction waste, which adds to the life-cycle inventory (LCI). Since there was no available data relevant to the local construction industry, surveys to collect data specific to local construction activities were conducted in this study. The estimated impact incurred during the local construction processes (e.g., those due to the use of various types of auxiliary materials, wastage) were then added to the cradle to entry gate data. Similarly, impacts that would arise at the life end of a building (mainly those incurred by landfill of the demolition wastes) were also included.

To reduce the amount of data that would need to be stored in the final database, instead of a set LCI data for each material, LCA was conducted using a detailed LCA software to yield an impact profile for each material, which comprises of 10 impact indicators, one for each of the impact categories embraced by the selected LCA method (CML 2 Baseline 2000). Only the impact profiles for various materials analysed were lodged in the database.

3 Exit gate to grave - besides the aforementioned impacts, LCA of a building would not be complete without accounting for the recurrent impacts that would be incurred during the operational life of a building. Such impacts include the replacement of worn-out materials and energy use for operating services systems in the building. The impacts due to the former can be forecasted on the basis of the life expectancy of individual types of materials and the per unit impacts, as quantified by impact profiles. The latter can be evaluated once the energy use throughout the operational life of the building is known, but this needs to be predicted using a building energy prediction model. The most sustainable design option can be identified after alternative designs are assessed. Thus, the model should be capable of providing energy use predictions efficiently.

7.4.3.2   Processes of developing the LCA tool

Establishing a localised LCI database in a key step for developing the LCA tool. Localisation of the LCI data retrieved from the propriety databases requires knowledge on the following in addition to the findings of a reasonably typical survey:

- The countries where building components and materials are imported from;
- The fuel mix for electricity generation in each of the country of origin from which the required building materials and components are imported;
- The mode and distance of transport from the country of origin to the project destination for each imported building component or material;
- The life expectance of building components, materials and services installations.

Additionally, the following information about local construction practices is essential to the localisation of LCI data:

- The type and amount of building components that are typically pre-fabricated off-site;
- The amount of materials that are consumed for temporary work;
- The amount of energy/fuel, water and auxiliary materials used for on-site construction of various components;
  The amount of construction waste generated.

Figure 7.5 shows a diagrammatic illustration of the localisation process, highlighting the type of additional data required for the process.

7.4.3.3   LCA software tools

LCA is a reliable means of gauging the sustainability of buildings. As a computer program incorporated with such philosophies, the LCA tool compiles environmental impacts and life-cycle costs of buildings and compares the impacts and costs for alternative designs.

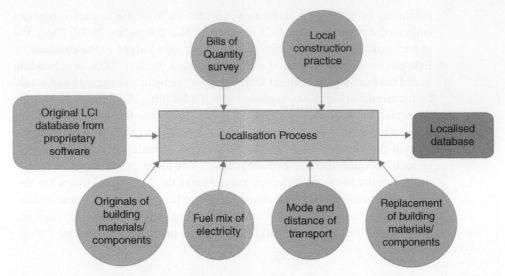

Figure 7.5   Localisation LCI Process Development. © Arup

The tool fits into the logic of a building designer by establishing a well-structured-data input system with (1) foundation, (2) floor, and (3) services potions. Each portion comprises a number of component and element groups supported by a range of libraries. Localised data impact profiles and weighting factors between these impacts were also built in for a user's reference.

Generally, the LCA tool conducts an analysis on the following aspects:

1 Life-cycle Impact Analysis - the tool compiles impact results for the various life-cycle stages of a building. The presentation of these results can be made into graphical forms for ease of comparison.
2 Life-cycle Cost Analysis - the tool calculates the life-cycle cost of a building with a lifespan of 50 years, which embraces the first investment cost and the recurrent cost of the building. The tool considers cost parameters like nominal interest, inflation rates and the price of electricity.
3 Operating Energy Model - an accurate and computationally efficient building model for predicting the operating energy use allows building designers to assess the energy consumption of various building service systems including HVAC, electrical, lighting, fire services, lift, escalators, and so on.

7.4.3.4   *LCI of buildings*

Since building involves a large number of various products and materials, LCA can be used to provide an objective assessment and allow selection to be made in the design stage for a more sustainable development. Through the suggested process to establish a database for a location, the projects within the area can make use of the LCI for the LCA. To put this concept into context, the following example is used to illustrate the impact of this tool (Example 7.4).

Example 7.4   LCA for Hong Kong

A study was conducted in Hong Kong to develop a LCA tool for commercial buildings. A building LCA model is usable (Figure E7.4.1) if it is supported by a database with impact data for a wide range of construction materials. In this study, a sample of representative commercial buildings was selected as a basis for identifying the most commonly used building and building services components and materials among commercial buildings in Hong Kong. To ensure the obtained data is timely and representative of local building construction practices and reasonably comprehensive, the sample buildings were selected from a random list of commercial buildings on the basis of the following criteria:

- The selected buildings should have been completed within the past ten years or were under construction during this study;
- Information and details for any selected building should be complete, readily available and accessible;
- The sample must represent a wide range of built forms, scale, grades, designs and specifications;
- The sample must embrace buildings equipped with different types of building services systems.

To facilitate the systematic organisation of a large amount of input data and calculations involved in modelling a complex building that may comprise of many components and services systems and equipment, a building is decomposed (conceptually) into a number of Portions. Each Portion is divided into Component Groups, which are further sub-divided into Components and Elements.

Additionally, four Government office buildings were included in the survey for comparison with commercial office buildings. In total, the sample included 28 buildings, among which there were 16 private office buildings, four government office buildings, four retail centres, and four hotels.

The impact points per square meter of Construction Floor Area (CFA) contributed by the building materials of the studied buildings were also determined (Table E7.4.1). The impact points per square metre (pt/m²) of CFA for a building range from 62 to 119. Concrete, rebar, structural steelwork, plasters/render/screed, and aluminium were ranked the most influential materials in terms of the overall impact points. These five types of materials and components combined would contribute up to 75% of the total impact points of an individual building. Concrete alone would contribute up to 31% of the total impact points for an individual building, despite that it would account for more than 77% of the total weight of building materials. On the contrary, rebar would contribute to more than 23% of the total impact points for a building, while it would only account for around 7% of the total weight of building materials.

Figure E7.4.1   LCA/LCC Modelling of a Building. © Arup

Table E7.4.1   Material List

Material Group	Total Weight per CFA $(kg/m^2)/(\%)$	Total Impact points per CFA $(pt/m^2)/(\%)$
Concrete	1468.76 (76.67%)	27.52 (31.26%)
Reinforcing Bar	137.29 (7.17%)	21.77 (23.76%)
Plaster, Render and Screed	95.18 (4.97%)	6.91 (7.39%)
Structural Steel	38.42 (2.01%)	6.11 (6.45%)
Bricks and Blocks	37.41 (1.95%)	5.27 (6.39%)
Structural Precast items	32.82 (1.71%)	3.91 (4.32%)
Access Floor Panel	21.19 (1.11%)	3.78 (3.97%)
Galvanised Steel	17.24 (0.90%)	3.34 (3.71%)
Glass	15.31 (0.80%)	2.46 (2.68%)
Stones	13.96 (0.73%)	2.39 (2.11%)
Aluminium	8.64 (0.45%)	1.74 (1.92%)
Tiles	7.68 (0.40%)	1.49 (1.62%)
Formwork	4.69 (0.24%)	1.40 (1.60%)
Stainless Steel	3.86 (0.20%)	1.23 (1.58%)
Thermal Insulation	2.57 (0.13%)	0.44 (0.43%)
Firestop Insulation	2.04 (0.11%)	0.15 (0.18%)
Plasterboard	1.24 (0.06%)	0.14 (0.16%)
Asphalt and Bitumen	1.11 (0.06%)	0.05 (0.05%)
Acoustic Insulation	0.89 (0.05%)	0.05 (0.05%)
Plastic, Rubber, Polymer	0.78 (0.04%)	0.04 (0.04%)
Fireproofing Coating	0.72 (0.04%)	0.04 (0.04%)
Chipboard	0.72 (0.04%)	0.03 (0.04%)
Acoustic Tiles	0.66 (0.03%)	0.03 (0.04%)
Cement Board	0.62 (0.03%)	0.01 (0.01%)

7.4.3.5   *Material use optimization*

Optimising the structural design of a building can reduce material use. Structural design evolves through an iterative cycle of proposals, testing and modifications. At an early stage, it is important to reason and justify the design concepts step by step, to understand the relationship between particular scenarios that a design creates and its responses to those scenarios. The objective of design justification is to build confidence in the solution, use methodology, methods, tools, and so on. A design engineer makes use of different methods to achieve this goal. The performance of a design can be predicted by prior knowledge, experience and forecast methods.

Justification often takes the form of an explanatory story that sometimes only exists in the mind of the design engineer. This case of justification explains how a structure works, and how it deals with different scenarios with unforeseen situations. Justification is based on gathered evidence and applying logic, making use of reasoning, thought, schematisation, calculations, modelling, analysis, and the development of scenarios. These justifications are based on a set of design assumptions and the definition of the requirements, design constraints and boundary conditions. Forming a conceptual story concludes the conceptual design stage and allows an engineer to present a generated solution to other parties in the design process or to the client.

In the present design practice, it is not common to use computational design tools to compose, define, explore, communicate and visualise structural design options. It is also uncommon to support an engineer with the formation of a conceptual story during early design stages. At a design stage, solutions of high-level accuracy are not required at this point. It is more important to compose and evaluate different options and obtain different scenarios in the impact of decisions. With the current tools available, it is difficult for engineers to compose and justify conceptual structural solutions, which contrasts with the often high-impact of decisions made during these early stages. Occasionally, it is only found out to be very costly at a later stage, and even impossible to compensate for design choices made at the beginning of the design process.

Another problem of the characteristics of computation and structural design is the black box problem. Software applications do not provide an insight into the internal working methods and it is therefore difficult for an engineer to completely rely on software. If an engineer does not comprehend the software or have confidence in its working methods, it will not be used. Another problem is that most software is not adaptable. It is not possible to adapt the software to suit particular functions or purposes in a simple manner.

Ongoing development of a design toolbox allows a structural design approach that combines parametric and associative modelling with direct analysis. This approach supports the appropriate speed and accuracy for the early design stages. The toolbox must assist the user in setting up the conceptual story as computable logic which can be used, re-used and changed in near real-time and to document this argumentation in a flexible form which can serve as a starting point for further design. This design story should always be capable of referring back to the concepts documented to check, control or verify the developments in the design against the documented benchmark formed by earlier reasoning.

Structural components provide a large number of analysis methods. The main aim is that these analyses are (near) real-time to streamline the sensitive and fast changing process of composition. The building components mentioned in the previous paragraph have structural design logic embedded in the form of simple rules of thumb, analytical and numerical analysis algorithms, standards, and so on. As the user assembles a design concept from the components mentioned above, the structural performance will be automatically analysed (Example 7.5).

Example 7.5   Zero material (material reuse) in Hong Kong

Environmental impacts from materials can be saved if we reuse most of them for new buildings. The Development Bureau of Hong Kong SAR Government looked for a temporary office accommodation (Figure E7.5.1) for the Energising Kowloon East Office (EKEO), which was set up to steer the transformation of Kowloon East. Upholding the Government's clear direction for sustainable development, especially for this new development area, it is important to stem a clear image of green building for EKEO.

A patch of unattractive land underneath the Kwun Tong Bypass was identified, which is ideally located at the heart of Kowloon East. After spending just six months from design to construction, a two-storey building that can accommodate 20 staff and 50 visitors within its 1,200 m² floor area was expeditiously built in May 2012. Apart from the required office facilities, there is an information kiosk with a briefing hall, exhibition area, two conference rooms, washrooms and a central courtyard.

The building adopted a number of sustainable design strategies, which include:

- Modular construction with recycled freight containers and mild steel framing structures fixed by bolts and nuts for flexible layout modification. These materials can easily be dismantled and reused elsewhere at the end of the building's life. Low embodied energy and sustainable materials were also applied.
- Passive designs that enhances energy footprint reduction of EKEO while creating a pleasant and comfortable environment. The Kwun Tong Bypass is used as shade for the building, which significantly cuts solar heat gain. Perforated external walls with cross-ventilated windows enhances cool breeze to circulate inside the office space.
- Energy efficient installations can reduce energy consumption significantly, such as variable refrigerant volume (VRV) air conditioners and T5 fluorescent tubes with daylight and motion sensors.
- Water saving measures can reduce water consumption significantly, for example low-flow taps and harvesting rainwater for irrigation.

Figure E7.5.1   Energizing Kowloon East Office, Hong Kong. © Arup

7.4.4   Operation stage – behavioural changes

During the operation stage of a building, the behaviours of two key players will determine the success of a sustainable building – the occupants and the facility manager (FM). Traditionally, these two parties do not often interact with each other. Only if a building is not functioning properly will they interact with each other. In fact, their collaboration can generate big potential for energy saving when running green buildings.

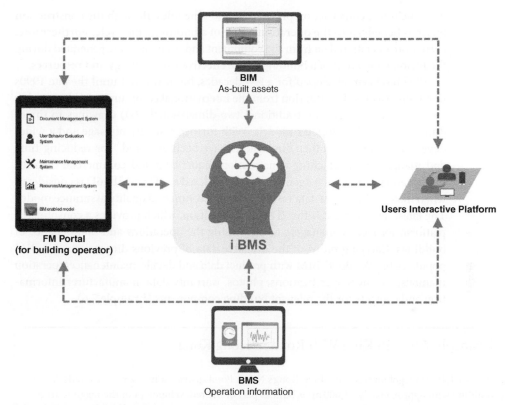

Figure 7.6 Integrated Building Management System. © Arup

To motivate occupants for behavioural changes, for example, using more daylight and natural ventilation, we should enhance the controllability of the individual occupants of their own spaces. Occupants should be provided with more information from the FM regarding the outdoor conditions for switching to an environmental mode of building operations.

For the FM, they should be provided with instant operational information of the systems and the performance of the space. The occupants with the help of IT systems can supply the latter efficiently. Access to information of indoor environments (being too cold, too hot, too bright or too stuffy), the FM can react instantly to fine tune the operational conditions of the systems (A/C and artificial lighting in particular) more efficiently. With the help of BIM information, the interaction between occupants and FM can be conducted effectively and make green building smarter (Figure 7.6).

7.5 BUILDING INFORMATION MODELLING

Building Information Modelling (BIM) is a platform that provides digital transformation for the building industry and enables building usage data to pass through different life-cycle phases of the development. The data generated, such as the dimensions of a building's features and systems and components sizes, during the design phase is used to the construction phase; and the additional data, including construction procedures and schedules, through the construction phase is beneficial to the operation phase of a building's life cycle. Furthermore, even more operation data from the equipment and occupancy are generated during the lifelong operation of a building to the effective use of energy and resources.

BIM has been developed for a few decades, but it was not until the late 1990s that there was a wider adoption from the international community. BIM started as an improvement from the traditional two-dimensional (2D) drawings to three-dimensional (3D) computer models. With further maturity of usage and a wider development of BIM, other dimensions have been explored. For reducing time and manpower by attaching construction sequencing and commercial data for project and portfolio management, it is considered as 4D and 5D. 6D is considered when material life-cycle data results in an environmental quality assurance model. The latest development is the 7D BIM application, which provides a consolidated platform for facility managers in optimising the operations and maintenance of buildings. Carrying on from the available data of previous dimensions, 7D BIM populates the "As-Built" BIM with product data and details, maintenance/operation manuals, cut sheet specifications, photos, warranty data, manufacturer information and contacts for the relevant building components (Example 7.6).

Example 7.6   18 King Wah Road in Hong Kong

In a revitalized neighbourhood in Hong Kong's North Point district, a new 25-floor Grade A office building with approximately 31,000 m² is built as part of the redevelopment of the neighbourhood. According to the vision of the developer, Henderson Land, the King Wah Road building (KWR) will be a sustainable and intelligent commercial building that can provide a healthy environment for users. The project utilised 7-dimensional building information modelling (7D BIM) throughout its design and construction stages, and carried the model through the operational stage.

Beyond the commonly used 3D BIM for building design, which can address "clash detection" between systems like building structure and equipment, the project used 4D BIM to address scheduling and planning. This is useful for construction sequencing and even maintenance procedures to be considered when design decisions are made. Time scheduling and planning could be optimised while the critical path is identified and addressed. The additional 5D dimension provides the cost aspect of the project. Not only can the cost impact of different design options be visualised, the building material life-cycle analysis is also incorporated. The next two dimensions are quite new relative to the current industry practice, where 6D covers sustainability parameters and 7D covers facility management during the building's operational phase. 6D assesses the environmental impact to people around the building based on design decisions. Based on design limitations and options, the resultant built environment, both outdoor and indoor, can be analysed; therefore, the health and comfort of the future occupants can be incorporated into the building design. As for the 7D BIM, facility management is incorporated. Even though the Building Management System (BMS) has been established for many years, it has only stayed at the operational level. The lighting, HVAC and other power-needed devices are connected to BMS for monitoring and controlling. What 7D BIM can offer is to bring in existing building information from previous design and construction processes to merge with the BMS infrastructure. Along with the increased connectivity made possible by the new technology of IoT (Internet of Things), big data can generate insights and actionable decisions for energy reduction and efficiency. The 7D BIM in KWR is implemented to create an office building that is sustainable, smart, and cost effective.

7.6   SUMMARY

Building a capacity of professional practitioners is of paramount importance if we want to see positive changes in the building industry transforming into the practice of green building. Some encouraging outcomes were observed in recent years on standardising the practice of sustainability in the building sector. International and local codes and standards were formulated to guide designs, construction and operations of buildings in the most sustainable manner. It is envisaged, through more collaboration amongst stakeholders, that these standards and practices will be improved to address the changing environmental conditions and social needs.

Chapter 8
Engineering solutions

8.1 INTRODUCTION

Planet Earth has been suffering from the negative impacts caused by climate change for many years. Our mission is to deliver sustainable solutions to avoid further damage to the environment and restore balance between human activities and our natural environment. However, this requires years of dedication and cross-generation cooperation. Meanwhile, our focus will also be placed on the resilience of our current buildings and infrastructure, which must be prepared to withstand extreme weather conditions.

The practice in sustainable development, particularly in building design, has equipped the practitioners with in-depth knowledge and wide exposure to assist countries in improving their resilience and adaptation to climate change. This experience is also important for the building industry to standardise practices in design, operation and the decommissioning of buildings. These mitigated measures need to apply at different levels of development, from city to district levels, and from building to component levels. For instance, in a city's master planning, drainage systems would need to handle more rain water flow with a higher frequency of occurrence. On a building level, structural integrity would need to withstand stronger winds and be resilient to earthquakes and fires.

Beyond raising standards for the basic required safety of a building's infrastructure, the building industry is finding new ways to enhance building performance, aiming to minimise negative impacts and even trying to achieve a neutral or positive impact. High-performance building is attempting this by using state-of-the-art technologies and strategies. Many design innovations will be showcased in this chapter. Each of these innovations not only allows buildings to serve their basic functions, but also enables buildings to excel in performance and other aspects.

Building Sustainability in East Asia: Policy, Design, and People, First Edition. Vincent S Cheng and Jimmy C Tong.
© 2017 John Wiley & Sons Ltd. Published 2017 by John Wiley & Sons Ltd.

8.2   DESIGN PROVISIONS FOR SUSTAINABLE BUILDING

Nowadays, sustainable buildings in Asia have to be designed with various provisions to address climate change and urban challenges. Three key considerations must be incorporated at an early design stage:

1 Resilience and adaptation to climate change or climate extreme;
2 High-performance to reduce carbon emissions;
3 Innovation to deliver quality indoor and outdoor environments.

For resilience and adaptation, buildings must be designed with top standards to withstand extreme weather conditions. Asia is highly susceptive to natural disasters such as typhoons, flooding/tsunamis, earthquakes, and fires, and so on. The structural elements and systems need to cater for the impact from these potential disasters.

Regarding high-performance design, sustainable buildings should be able to provide a comfortable and healthy environment to safeguard the well-being of its occupants on one hand, and consume the least amount of natural resources such as water and energy on the other hand. Designers are tasked to deliver such requirements by means of passive and active design.

On the topic of innovations, designers are required to stretch the limit of conventional technologies and incorporate the breakthroughs on the sciences of materials, physics, biology, and telecommunications, and so on, of building designs. These aspects in particular can challenge our practice from time to time.

8.3   ADAPTATION TO CLIMATE CHANGE AND RESILIENT DESIGNS

The structural design of a building is fundamental for a number of reasons. Past experiences show us that buildings have to be rigid and provide integrity to be sustained throughout their lifespan, that is, 50 years. However, from a sustainability perspective, it is also important to provide flexibility during 50 years of operation, which is based on changes, usage, and operating conditions. As covered in Chapter 3, the negative effects of climate change or climate extreme have surfaced in recent years. It is becoming apparent that more and more extreme weather will be occurring in the future. In the Asia pacific region, severe typhoons and tsunamis caused by earthquakes will become more frequent. It is therefore imperative for buildings to be adaptive and resilient when we discuss the sustainability of buildings.

8.3.1   Extreme wind engineering

In the design of tall buildings, the prediction of dynamic response subjected to strong wind associated with extreme weather plays an important role.[107] The loss to life and property can be huge. Cities such as Hong Kong, where typhoons are common, has made an effort to develop wind codes to guide the design of buildings (see Example 8.1). The code itself needs frequent review in light of the worsening wind conditions caused by climate extreme.

Example 8.1   Typhoon model in Hong Kong, China

Little research has been conducted on full-scale validation of the empirical modal PSD obtained from wind tunnel tests. The lack of success is mainly due to the feature of modal PSD being an integrated quantity over a building body. In addition, it is not a quantity that can be directly read from measurement or determined by integrating local pressures, as it requires high frequency wind pressure exerted on many planes on the building.

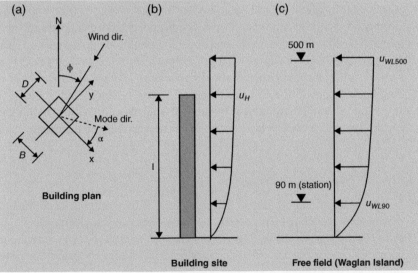

Figure E8.1.1   Typhoon Satellite Image and Building and Wind Profile. © Arup

It remains as a challenge to derive the modal properties (including the modal PSD of loading) from stochastic "output-only" response. Conventional methods do not seem to be sufficiently accurate by giving inconsistent results with wind tunnel tests. Recently, an algorithm based on fast Bayesian Fourier Transform (FFT) modal identification has been developed, aiming to correlate the calculation modal PSD of wind load for tall buildings to full-scale field observations. The method allowed the prediction of modal PSD with the aid of wind tunnel tests and mean wind data at the Hong Kong Observatory (HKO) Waglan Island Station (Figures E8.1.1 and E8.1.2). By using the data at the corresponding instant followed by the Bayesian identification method, the wind tunnel results were correlated to field data taken during the strong wind events. So far this method seems to be a reasonable analytic tool as it has demonstrated fair agreement between wind tunnel results and field measurements.

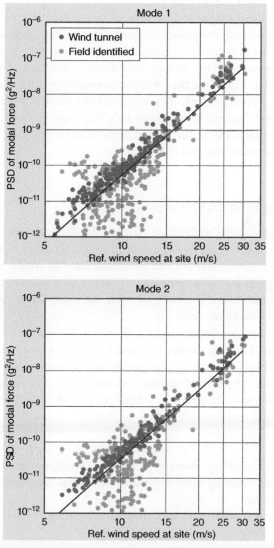

Figure E8.1.2 Comparison between Wind Tunnel Prediction and Field Identified PSD. © Arup

The dynamic structural response under gusty wind loading is a combination of quasi-statics and stochastic dynamics. The quasi-static response is primarily dependent on two factors: the mean wind level and a building's lateral stiffness. The dynamic response can be divided into two components: background (depending on overall energy in the wind spectrum) and resonance (depending on a building's modal properties including natural frequency, damping ratio, mode shape, modal mass and modal power spectral density (PSD) of dynamic wind load. In particular, PSD of dynamic wind load is an important factor affecting the dynamic response of tall buildings under strong wind as it represents an effective loading quantity, integrating a range of wind load characteristics. Generally, the variance of dynamic response is proportional to the modal PSD and its value can change by orders of magnitude under strong wind conditions.

Many engineering models have been built attempting to calculate the modal PSD for structural design applications. The models consider accepted scaling models of mean wind distribution, turbulence intensity and spatial variability. For structures prone to vibration, wind tunnel tests can be used as an effective experiment to determine the wind forces employing on a scaled-down model in a controlled environment. By correlating the results of wind tunnel tests and realistic field measurements taken in a full-scale building, the practicality of wind tunnel tests can be further strengthened as it generates critical information to assess the actual risk and conservatism.

8.3.2   Flood mitigation and prevention

With rising sea levels and climate extreme, flooding will occur more often in locations prone to this risk. These locations often take place near the coast and rivers, where they face flooding from oceans, lakes or stream networks. This type of flooding can be characterised as still water or velocity events where either no horizontal movement is associated with the rising water or fast moving water found in any depth or water.

In order to prevent or reduce flooding damage to buildings near flood-prone areas, building designers should conduct a thorough evaluation of the expected flood levels for the design lifetime. Design guidelines are available in most locations; however, frequent updates are needed as climate extreme affects the conditions more often. If a flood risk is identified, it is still possible to include flood mitigation measures to lower the effect from floods; some of the methods are as follows:

- Dry flood-proofing - prevent water entry by making the building watertight;
- Wet flood-proofing - increase water resistant to inhabited and critical parts of the building;
- Floodwalls - keep water away from the building by incorporating floodwalls into site design;
- Relocation of the building - if there are no other mitigations, it is not a good idea to have the building in such a location.

Floods are not only caused by climate extreme events, it can also due to an extreme event like a tsunami. A tsunami is caused by seismic activities in the oceanic plateaus, landslides or volcanoes. For instance, a tsunami wave, which is caused by an earthquake in the ocean, may start with a small surface magnitude, but may travel to shore with a fast moving wall of turbulent water that can cause catastrophic damage. Due to the impact of such an event, many building guidelines have started to include new sections to cover tsunami loads and effects, like ASCE7-16.[108] The aim is to minimise the impact through urban planning, relocating away from shorelines, strengthening breakwater structures, preparing community awareness, and developing timely warnings and effective responses.

8.3.3   Seismic design

Many cities in Asia are located in regions prone to earthquakes. Stringent seismic codes are in place, yet the most recent devastating earthquake in Sichuan, China has provided engineers with a need to look into the practice in greater detail.

Over the past decade, reinforced concrete walls have been widely used as a seismic resisting system, especially in high-rise buildings. The behaviour of such buildings in design level events is relatively unknown, as most major urban centres have been spared from large magnitude earthquakes. However, data from the recent 2010 earthquake in Chile confirmed that reinforced concrete walls are vulnerable to significant structural damage, and in some cases collapse.

Seismic design has often been overlooked in many national building codes where only minimum seismic requirements are imposed. For instance in the US, building codes require buildings to remain structurally undamaged in relatively frequent events and to provide life-safety performance for much larger ones (accept considerable and irreparable damage to structural and non-structural components).

To ensure building safety performance in seismic events, non-linear computational analysis can be used to predict a building's response. The analysis incorporates 3D numerical representations of a building to historic ground motions and can be adopted to meet project-specific objectives or to enhance seismic performance. This tool helps structural engineers to deliver safer designs. However, this analysis still relies on computer capability and reliability.

Validation is a key issue due to the lack of data from concrete wall buildings in earthquake events. Structural engineers and researchers must compare their computational results against physical tests carried out in laboratories. Normally, tests are conducted under a quasi-static state, which involves specimens being loaded cyclically and very slowly in each direction. The amplitude is increased for each cycle until the specimen fails. Unfortunately, such tests have their limitations – given the fixed loading protocols and the single mode behaviour, the tests fail to take into account the effects of damping and stiffness degradation on realistic demands. Therefore, large-scale dynamic shake table tests are required to overcome these restrictions.

In 2006, a cornerstone was marked at the University of California, San Diego where a full-scale 7-storey reinforced concrete shear wall assembly was constructed on a large outdoor shake table. A suite of ground motions with increasing intensity – including one of the strongest recorded in the 1994 Northridge California earthquake – was applied on the wall assembly. This test, being the first of its kind around the world, generated comprehensive empirical data of a reinforced concrete wall experiencing actual nonlinear behaviour. A blind analysis competition was organised to allow contestants to predict the structure's response. Eventually, no team was successful in giving an accurate prediction of the results. It was then concluded that no software on the market was powerful enough to precisely model actual seismic behaviour of a concrete wall structure.

8.3.4   Fire engineering

Fire protection is amongst the most basic provisions of safeguarding a building and its inhabitants. The collapse of the World Trade Centre in 2001 has particularly galvanised the community in the US, with the output from the WTC investigation very much tending towards serious code changes to allow structural fire engineering. Structural fire engineering assesses the real structural response of heat and uses this as a basis for fire protection requirements and structural detailing to improve robustness. In its simplest form, it can often allow a reduction in overall building ratings or justification for specific localised member designs that may fall outside an overall code-based approach. For high-risk or super high-rise buildings, it allows engineers to move away from very simple code requirements for fire ratings, and to provide technically sound and robust designs that can cater for the fires possible within their building. In its most complex form, it can be used to justify leaving all secondary beams unprotected, global reduction in fire proofing thickness or concrete cover requirements, plus creating specific detailing to improve stability and/or compartmentation requirements with less reliance on applied fire proofing.

The performance-based design methodology allows fire resistance requirements to be derived that address the actual hazards in buildings. This process can result in reductions, and in some cases, elimination of fire protection to structural elements for particular occupancy types and building shapes, or increased ratings for high-risk buildings such as multi-level warehouses. The objective of performance-based engineering in this case is to satisfy the life safety requirements of building regulations by providing a level of protection to structures that are required to survive a fully developed fire scenario in a building.

8.4   HIGH-PERFORMANCE BUILDINGS

With the basic aspects, which are mostly safety related, being accounted for in the design process, many other considerations are to be included as the next integrated step towards the development of sustainable buildings. High-performance building (HPB) is a prevailing concept for building design.[109] Designers worldwide are applying various methodologies to achieve a performance standard.[110,111]

The performance could be a requirement prescribed by the client or the ultimate goal of zero carbon emissions. The recent development on computational modelling has supported a lot of analysis on performance evaluation. It is now a standard practice for designers to carry out simulations to verify the effectiveness of a design. Performance-based evaluation is becoming the norm for design standards and regulations alike. Performance criteria, such as thermal comfort, energy, ventilation, daylight, air quality, and so on, can be readily quantified such that the options of design can be compared and optimised. Example 8.2 demonstrates the result of such a design process.

Example 8.2   Hysan place: Integration of commercial necessity and sustainability in Hong Kong, China

Figure E8.2.1   Hysan Place in Hong Kong. © Kenny Ip

Even with environmental constraints from a high-density urban area, Arup designed an integrated sustainable strategy for this high-rise building, which not only enhances the building's performance, but also benefits the local community with its access to a green public space. Hysan Place (Figure E8.2.1) was redeveloped on the site of the former Hennessy Centre, and it faced several environmental and social challenges due to its surroundings. Located in one of Hong Kong's most vibrant commercial districts, Causeway Bay, people enjoy exceptional transport and other conveniences. That being said, they also suffer from some of the highest air pollution levels, heat island effects, and lack of open space. This redevelopment implemented a tailor-made building sustainability solution for high-density urban areas.

The design strategy considered beyond energy-efficient systems and the use of low-emissions, and recycled materials. Porosities were also given to wind and orientation for solar to maximise daylight and natural ventilation, as well as internal planning flexibility for long-term re-use. The building boasts a series of green features including light shelves, solar shading, low emissivity glazing and vents, a green roof, and a native plant sky garden for the public's enjoyment. In addition, the following are the highlights of the building's design:

- Urban windows - large openings at lower levels in the building were purposely designed to allow airflow through the area and reduce wall effects to the street level. This feature enhances air ventilation in the neighbourhood, and hence, air pollution can be reduced. They also provide sky gardens that can reduce heat island effects.
- High performance envelope - to cool and illuminate the indoor space effectively, the façade was designed to allow natural ventilation and lighting. To promote natural ventilation, local stack effects were fostered by lower and upper vent designs. To improve the uniformity of daylight and energy efficiency, large window areas with shading and light shelves were used to balance the amount of daylight.
- Advanced ventilation - to maximise the energy benefits and thermal comfort for the occupants, both natural and mechanical ventilations and air-conditioning were carefully programmed during the transition seasons or non-office hours.

As a result, the development has reduced portable water use and carbon emissions significantly compared with existing typical retail and office mixed-use buildings. The strategy has reduced air pollution and the heat island effect, and has allowed an open space for the local community. Hysan Place showcases the possibility for greener and better development; and the project achieved BEAM Plus Platinum Certification and LEED-CS Platinum Certification.

8.4.1   Building physics analysis

Building physics is the application of the principles of physics to improve built environments and it is a tool to assist building HPB. It brings a fundamental understanding of physics to improving the design of building fabrics and surrounding spaces. A range of advanced analytical tools, skills, and techniques allow us to work with clients to design buildings that are comfortable to occupy, easy to use, and light in terms of their environmental impact. Using advanced methods of analysis in combination with design creativity, buildings can respond well to the climatic conditions of their chosen site, function efficiently, are pleasant to occupy, and hence, are economic to run.

Building physics techniques allow us to gain an in-depth understanding of the environment and physical properties of materials that affect buildings.

By modelling current and future performance, we are able to address the challenges to building designs posed by factors such as internal and external airflow, condensation, heating and ventilation, energy use and artificial/natural lighting and shading. By using building physics, practitioners can deliver design solutions that ensure that buildings are sustainable and energy-efficient, subject to reduced risk, more usable and more comfortable.

Building engineering physics addresses several different areas in building performance including: air movement, thermal performance, moisture control, ambient energy, acoustics, light, climate, and biology. This field employs creative ways of manipulating these principal aspects of a building's indoor and outdoor environments so that a more eco-friendly standard of living can be obtained. Building engineering physics is unique from other established applied sciences or engineering professions as it combines the sciences of architecture, engineering, and human biology and physiology. Building engineering physics not only addresses energy efficiency and building sustainability, but also a building's internal environment conditions that affect the comfort and performance levels of its occupants.

8.4.2   Energy appraisal

High performance, in terms of energy usage, can be demonstrated by conducting an energy appraisal. Whole building energy simulations are required at the early design stage to predict the energy consumption of a building when it is built. It is a very powerful tool that generates the required information for making decisions when design options are proposed. The results were used to size the building's energy systems as well as renewable energy systems for neutralisation. At a design stage, this process was also important to evaluate the effectiveness of different design strategies to reduce the loads (cooling and electricity) and energy demand. This helped architects and engineers make design decisions using the standard cost-benefit analysis.

Building energy simulation was carried out using hourly weather data.[112] The building design, including architectural elements and building systems, were input into a simulation tool. It adopted the typical meteorological year (TMY) method,[113] which was developed by the Sandia National Laboratories in the US and had been commonly used in the past. A TMY comprises of 12 typical meteorological months selected from several calendar months in a 25-year period (1979–2003) weather database.[114] The Hong Kong prevailing climatic characteristics were represented by an 8,760 hourly TMY weather file.

8.4.3   Indoor environment quality

Another aspect to quantify an HPB is from the Indoor Environmental Quality (IEQ). IEQ is imperative for the design of sustainability as city dwellers spend 90% of their time indoors. It safeguards the health and comfort requirements for works and habitation. A good IEQ design can improve the quality of living and wellbeing of inhabitants. The quality of IEQ is defined by a number of performance indicators. The most prominent include Indoor Air Quality (IAQ), daylight, thermal comfort, and noise.

8.4.3.1   Indoor air quality

IAQ is becoming very important for non-domestic buildings, such as offices, hotels, retail, hospitals, and so on, nowadays as most of these buildings are now designed with centralised air-conditioning systems. Sick-building Syndrome, which was once a big health concern in the 1970s, is a key design consideration to prevent this type of occurrence from happening and safeguarding the health of occupants. Occasional incidents of Legionnaire disease have still been occurring in the region recently. Design guidelines are in place to advise good practices of IAQ design. Guidelines of the design of cooling towers and the fresh supplying system of HVAC are amongst them.

From a green and sustainable building perspective, IAQ has other considerations other than purely health. Studies have found that there is a close relationship between good IAQ and productivity. The indirect benefits of good IAQ are enormous and becoming a key value for properties.

8.4.3.2   Daylight

Daylight is an important factor, both at work and at home. Studies show that daylight environments can enhance work efficiency. People tend to feel happier with day-lit environment. It can also create a positive psychological response to some specialist buildings. Patients in a well day-lit hospital have the potential to recover faster.

The quality of daylight is not only about its amount but also its distribution. A modified design will have very high lighting levels close to the window and decay quickly towards the interior space. This will not only reduce an area's feasibility for daylight, but may result in poor lighting quality resulting in a glary space. An improved design will incorporate a light reflector or light pipe to redistribute the available daylight.

8.4.3.3   Thermal comfort

A comfortable space can enhance a user's satisfaction and therefore, their overall productivity. Thermal comfort of a space is determined by a number of factors, including environmental conditions such as air temperature, relative humidity, and air movement, and also the behaviour of occupants such as metabolic activity and clothing. For most office spaces, air conditioning is designed to control the temperature and humidity of air to achieve the thermal comfort requirement. This involves the use of significant amounts of energy. More advanced A/C systems such as radiant cooling or desiccant dehumidification were developed to reduce energy use. The design of spaces with natural ventilation will have to consider the accommodation effect of occupants. This is particularly applied for residential developments. Thermal comfort curves for different climates were developed to assist in the design of natural ventilation.

Data from Post-Occupancy Evaluation (POE) can provide feedback of the occupants, which is valuable on the design of re-calibration of the system operation. By taking into account the "as-built" conditions and the feedback from the people operating and using the space, the operating parameters of system and equipment can be tuned to reflect the actual operation and further be optimised for their performance.

8.4.3.4   *Noise*

Noise is a significant issue in any high-density built environment, like in Mumbai, Tokyo, Shanghai, Hanoi, Manila, to name a few. High and low-frequency noise causes human health and discomfort. According to World Health Organisation (WHO), noise pollution can be the cause for ischemic heart disease, cognitive impairment of children, sleep disturbance, tinnitus, and annoyance. For instance, traffic noise is the major concern, but proper planning can generally deal with this problem. For congested sites, the conflicts between provisions of lighting and ventilation and the noise abatement require a concerted effort for the design team for an integrated design. By separating the source of noise and the receivers from people dwelling from any given developmental constraints, the noise impact to people can be minimised.

8.4.4   Outdoor environment quality

From an indoor environment for the direct experience of occupants, an outdoor environment also plays a role in the design process. Sustainable design should start from the early planning stages of the development. The objectives are to ensure the most comfortable microclimate induced by a new development. This can modify or improve outdoor comfort in a built environment and have minimum impact on its surroundings. A better microclimate can also facilitate a better utilisation of building passive designs. For example, cooler outdoor air can increase the chance for natural ventilation.

There are a number of environmental parameters that influence a microclimate and outdoor comfort quality. Those parameters include wind environment, solar environment, greenery, water features, outdoor paving materials and thermal comfort. Focusing on each individual parameter above will not effectively assist a project team to achieve the best design. Therefore, a reliable parametric study is required to integrate various parameters in order to create an environmentally optimised solution. In subtropical cities such as Hong Kong, solar radiation, wind speeds, and temperatures have the most influence on how a person feels thermally. A parametric study is able to provide an optimised solution in terms of thermal sensation. The optimised solution will then form a part of an integrated building design.

8.5   DESIGN INNOVATIONS

Once the basic requirements have been met by the building design, high-performance building is set to find ways of neutralising building-related carbon generation or even providing a positive contribution to communities. The following section showcases some innovations in building design. There are some designs that focus on the application of the building's facade, whereas others are solutions for allowing low energy performance for the building. These innovations deliver the basic requirements for safety and comfort for occupants, but they can also allow buildings to operate with low energy, hence, a lower carbon footprint. Some buildings might even generate renewable energy that can offset carbon emissions by the community. Example 8.3 illustrates how design innovations are put together in a state-of-the-art building project.

Example 8.3   Parkview green in Beijing, China

Parkview Green (Figure E8.3.1) is an innovative mixed-use building structure enclosed by a glass and ETFE plastic glazed envelope that houses offices, a hotel, and retail spaces. The envelope creates a microclimate environment with relatively uniform and easily controlled air buffer zones. The buffer zones enhance thermal insulation in Beijing's harsh winter, whereas ventilation louvers facilitate the hot air run-away from the envelope during hot summers. As hot air escapes, cooler air is induced from the bottom entrances of the buildings, and thus natural ventilation is generated.

Eco-shelter - Parkview Green features a total of four towers, including premium-grade offices, a six-star hotel and four floors of retail space, inside a microclimatic enclosure. It was challenging to find glass-like building material light enough to line the sloping roofs with; ETFE (ethy ltetra fluoro ethylene) was selected, which is not only light-weight, but also offers high-corrosion resistance and strength over a wide temperature range. The building utilised the principle of "through-air" using a stack effect, which is naturally ventilated by automatically opening and closing panels in the facade. Air-ducts were laid three metres underground, taking air in from the street level. The soil temperature at 15-17 °C all year round cools the air in summer and heats it in winter, leveraging on greenhouse effects.

Green accomplishments - with a naturally temperate indoor environment with a reduced need for heating and cooling, the project was enabled to cut energy use by 50%, compared to other buildings of a similar size. It is one of the world's most sustainable architectural developments, and achieved a Platinum level LEED rating (Leadership in Energy and Environmental Design). Besides, its visitors and occupants are protected from air pollution in the open, which can reach severely high levels. Parkview Green also provides refuge from sandstorms as well as the extremities of Beijing weather.

Figure E8.3.1   Parkview Green in Beijing, China. © Parkview Green

Anti-quake designs - the project eschewed the design of rigid structures, which risk falling apart during sideways movements caused by earthquakes. The different building parts, such as the glass roofs and walls, were fitted in a way that provides leeway among the contact points and adjoining sections. This allows the seismic shocks to be absorbed as well as transferred from one structural component to another. A 235-metre cable suspended footbridge, the longest in Asia, inside the complex displays the same design ingenuity. Both ends of the bridge were integrated into two buildings overlooking each other to give it robust structural support. This helps to absorb quake tremors and minimise the likelihood of the bridge's collapse during an earthquake. The bridge is also an eye-catching feature of the building, and forms an interesting entry/exit point for users of the building.

Fire protection - fire rescue, escape, and evacuation were also important concerns. The glass and ETFE envelope brought many fire safety design challenges and a "melting system" was introduced to the ETFE top, where it had to be able to melt during a fire, and to make the top open for natural ventilation. A temporary safety area was created for occupants and shoppers. It has an area between the high-rise buildings under the envelope, and the fire risk is low and people can leave the building from this area. A special path was also created for fire trucks to directly enter the complex.

Green pyramid - with its stunning facade, sustainable design, and risk-proof facilities in place, Parkview Green has set new industry benchmarks for future buildings, while also enabling those who live, work or visit there to experience an unrivalled level of comfort and safety.

8.5.1   Outside building: High-performance envelope

High-performance, low-energy façades actively recognise and optimise synergistic impacts on lighting, ventilation, and air-conditioning (HVAC) energy use, achieving greater energy-efficiency, comfort, and amenity compared to conventional piece-meal solutions (Figure 8.1).

A parametric study of design strategies will provide design information to building envelope details by considering the facade solar heat gain, orientations, sun shadowing conditions and various shading device effects.

The above figure shows the general process for envelope optimisation. The variation of solar heat gain on the facades will give rise to different window size patterns. The facade area where heat gain is higher will have a reduction in window size in order to decrease solar heat gain. Moreover, shading size and orientation can also be optimised based on the sun's position in a specific location. This comprehensive parametric study will finally generate an optimised envelope design.

With various facade patterns suggested, architects can use this information for their envelope design. Architects can selectively choose some of the facade characteristics such as window size design and shading patterns. After that, these features can be integrated into the facade and form part of the whole building's outlook design.

8.5.1.1   *Environmental responsive for extreme weather*

A façade acts as the skin of a building. It should own the function of protecting the building from the adverse (and changing) external conditions by moderating extreme conditions.

Integrated facade design

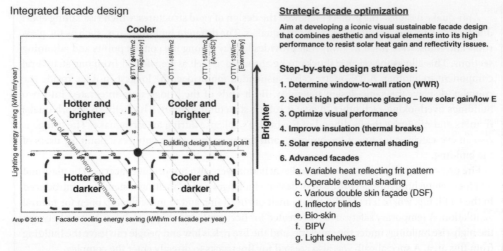

Arup © 2012   Facade cooling energy saving (kWh/m of facade per year)

Strategic facade optimization

Aim at developing a iconic visual sustainable facade design that combines aesthetic and visual elements into its high performance to resist solar heat gain and reflectivity issues.

Step-by-step design strategies:

1. Determine window-to-wall ration (WWR)
2. Select high performance glazing – low solar gain/low E
3. Optimize visual performance
4. Improve insulation (thermal breaks)
5. Solar responsive external shading
6. Advanced facades
   a. Variable heat reflecting frit pattern
   b. Operable external shading
   c. Various double skin façade (DSF)
   d. Inflector blinds
   e. Bio-skin
   f. BIPV
   g. Light shelves

Figure 8.1   General Process for Building Envelope Optimisation. © Arup

A typical building is designed with walls facing in different orientations. These walls are susceptive to different conditions at different times. A good design must consider such an important variation, in particular for a building design with a passive design. This can also avoid any unnecessary wastage of materials due to being over-provided by averaging the design conditions. In the context of hot and humid climates, solar protection, by means of shading provides a valuable opportunity to optimise the design the provision of daylight and the screening of solar heat. Due to different magnitudes in solar irradiance for the four elevations, the North elevation receives least solar irradiance and therefore has the advantage to act as a daylight source, and shading should never be applied. On the other hand, West, South and East façades should provide shading to minimise solar effects. Due to solar angle, vertical shade should be provided for West and East façades, while horizontal shading should be applied to the Southern façade.

The design should also consider the annual effects of the weather. In Northern parts of China, heating provision is dominant annually. An over-design for shading may result in higher energy consumption and unfavourable interior thermal comfort conditions due to the obstruction of solar penetration during winter. The depth of shading then becomes another critical design parameter that must be optimised. If glazing properties are considered, the processes of design optimisation will become prohibitively impractical when doing it manually. Nowadays, the building designers rely on computers to conduct parametric studies on different design options. An optimum point can be determined readily for controlling design parameters such as window size, glazing properties, and shading design for different orientations (Figure 8.2).

The design should also consider extreme climatic conditions in different locations, especially in the context of Asia. Climatic challenges include extreme temperatures (and its variations), humidity, and natural disasters, including sand storms in the Middle East and Northern parts of China, and typhoons in the Western coast of the Pacific. The design needs to maintain its integrity and function under such extreme conditions. Even under normal design conditions,

Annual cumulative irradiance

Shading mask - grey area indicates shading effect on facade, the more grey area, the lower solar heat gain

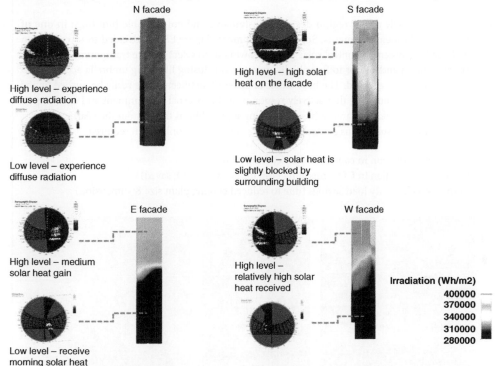

Figure 8.2   Example of Solar Responsive Shading Analysis. © Arup

the façade design has to make provisions to enhance its maintainability and serviceability. Example 8.4 illustrates an innovative facade design for the Al Bahar Towers in Abu Dhabi. These provisions are important from the life-cycle perspective of a building.

8.5.1.2   Bio-reactive façade

Zero-energy and surplus-energy buildings are thought to be the major focus of the sustainable architecture industry. To attain a positive energy balance, the building is not only required to operate with minimum energy, but also maximise the generation of on-site renewable energy. For now, photovoltaic (PV) and solar thermal collectors are examples of the more mature building-compatible renewable energy technologies. However, the yield of these two sources on a building scale is limited. On the other hand, biomass being the largest source of renewable energy has received little technological support to make it incorporated in buildings.

Fortunately, a facade system integrating flat panel photo bioreactors (PBR) as shading devices was developed. PBR are transparent containers, which provide a controlled environment for fast-growing species, such as microalgae, to carry out photosynthesis. The microalgae circulate with water and gain nutrition through

Example 8.4   Al Bahar Towers in Abu Dhabi

For centuries, people of the region designed sustainable and comfortable buildings in one of the world's harshest environments. So, the development (Figure E8.4.1) decided to introduce an external shading system to control solar gains, allowing us to select a more transparent glass. This idea of triangular panels, floor to floor high, opening and closing like a big umbrella according to the sun's position was adopted. The boundaries of adaptive architecture are really be pushed by the industry, designing a building that actively responds to the external environment and modifies its performance, shape and appearance based on changing conditions throughout the year.

Benefits: Effective solar gains control – reduced g-value (0.08) only when required:

- Approx. 15% reduction in capital cost for cooling plant
- Approx. 20% reduction in CO_2 emission (1140 tonnes/yr. of CO_2 saved)
- Approx. 20% electricity load savings (due to reduced cooling plant size & dimension)

Figure E8.4.1   Responsive Façade at Al Bahar Towers in Abu Dhabi. © Arup

panels where light penetrates through and initiates photosynthesis. The microalgae absorbs carbon and produces biomass and solar thermal heat. The PBR are connected to a plant room where carbon released from the combustion process in the neighbourhood feeds into. Eventually, the algae can be harvested and converted into methane on- or off-site. The heat is withdrawn from the system by heat exchangers, then either stored geothermally or fed back into the building for heating and hot water supplied by heat pumps. Example 8.5 summarises the first building integrated PBR systems in Germany.

8.5.1.3   *Zero energy wall*

Zero energy wall is an organic system at the skin of a building that absorbs energy from the façade and uses that energy for the activity in the wall. As a demonstration project, a self-sufficient system utilises PV to capture the energy during

Example 8.5   BIQ house in Germany

To explore the potentials and benefits of PBR, the following innovations (SolarLeaf) have been introduced by Arup, Strategic Science Consult of Germany (SCC) and Colt International when PBR were used on the building skin of Bio-Intelligent Quotient (BIQ) house (Figure E8.5.1), and this project was supported by Zukunft Bau and Splitterwerk Architects:

- A multi-functional facade component
- Full integration into the energy concept
- The system generated high-value biomass that potentially can be used by pharmaceutics and food industry
- Carbon is absorbed at the source of domestic and industrial sources
- Vibrant and refreshing architectural expression

Figure E8.5.1   BIQ House in Germany. © Colt/Arup/SCC

the day-time and store energy for the usage of LED (light emitting diodes) displays during the night-time. The idea of this innovation is to incorporate the sustainable technology as new architecture façade integration and to show the economic viability of such systems in building design. The density of the PV cells patterns is carefully determined and laminated within the glass curtain wall. The changing density of the entire building façade allows not only harvesting energy effectively, but also balances the natural lighting to get into the building's interior to save the cost of artificial lighting as well as reduce the heat gain into the building to save the cost of air conditioning. The size of the LED display is carefully matched between the energy use to illuminate the screen after dark and the energy gained from the sun when it is shining. Net zero energy allows a natural organic system to sustain the operation of the wall by itself. One project that incorporated the zero energy wall idea can be found in Example 8.6.

Example 8.6   Zero energy media wall in Beijing, China

Zero Energy Media Wall in Beijing is one of the largest in the world, and unique in being wholly self-sufficient in energy. GreenPix operates an organic system that harvests solar energy in the day, and uses it to power illuminations at night. The media wall harvests more energy than it consumes, so any daily excess is channelled into the national electricity grid. GreenPix transforms the building envelope into a self-sufficient organic system, harvesting solar energy by day and using it to illuminate the screen after dark. Arup provided lighting design and façade engineering for the project.

Using advanced computer modelling and analysis, our lighting and façade engineers used a combination of LED nodes spaced behind the glass exterior to create an indeterminate depth that is interesting to the eye in the daytime. With advanced computer software controlling the LEDs, the media wall comes alive in the evening.

This media wall behaves like an organic system, harvesting the energy it consumes illuminated at night. The integrated design team has developed a new glass façade where polycrystalline photovoltaic cells are laminated in the glass curtain wall. This is the first time such a system has been adopted in China, and with great success.

The lighting engineers conducted lighting simulations and various prototype studies to gauge the effects of 2,292 LEDs. The LEDs are controlled by embedded custom-designed software. The result is a spectacular "intelligent skin" that displays art-specific visual communications.

8.5.1.4 *Hybrid ventilation*

The provision of air-conditioning to meet particular comfort requirements requires a lot of energy. The maintenance of the equipment is also significant during the life cycle of the building. Hybrid ventilation is an enhanced design that can make use of the favourable outdoor conditions during mid-season to provide heating and cooling. To achieve this, the A/C system should be designed to be operated in three modes: natural ventilation, free cooling, and fully A/C. The design of the building and the A/C system should be enhanced accordingly. The operation of the hybrid ventilation is as follows:

- Natural ventilation mode - during favourable outdoor conditions, the window will be opened to allow sufficient air movement for ventilation purposes. No equipment is in operation.
- Free cooling mode - during mid-season, fresh air is drawn into the interior for cooling purposes by means of mechanical ventilation. Since the chiller is not required in operation, significant amount of energy can be saved.
- A/C mode - the A/C system will be operated at its design conditions.

Depending on the weather, hybrid ventilation can save up to 10% of energy for HVAC. Locations with longer shoulder seasons generally perform better as the HVAC can operate for a longer period under the natural ventilation or free cooling mode.

The building's control of the hybrid ventilation system is slightly complicated, as the control mechanism has to be coupled with outdoor conditions. In operation, the weather sensors will detect the outdoor temperature, humidity, as well as air quality to determine if natural ventilation or free cooling should be operated. Aesthetically, the openings can be integrated with elevation design consistently where a selection of actuators are available.

8.5.1.5   Acoustic window

As mentioned earlier in this chapter, traffic noise is a significant problem when designing residential buildings in densely populated Asian cities. Accurate acoustic performance modelling for facades is becoming increasingly important, especially during the current widespread adoption of natural ventilation as a sustainable strategy. Generally, it is thought that natural ventilation can improve indoor air quality by diluting indoor air pollutants, promoting thermal comfort, reducing energy consumed by air conditioning systems, and less space requirements for mechanical plants.

Unfortunately, natural ventilation is often forced into the compromise of external noise penetration through facade windows. Usually, external noise is either too high to be reduced to the recommended levels despite practical measures, or the cost of mitigation outweighs the benefits. Therefore, the application of natural ventilation becomes very limited.

Acoustics consultants also face another problem – to provide reliable predictions of sound insulation performance of naturally ventilated facades. Even though the glazing and internal partition sound insulation performance is widely available, and little is known about the different configurations of operable windows and ventilated facades. Specific to this problem, the Ventilated Acoustic Cavity Solver (VACS) can be used to efficiently estimate acoustic performance of various configurations of ventilated double-glazing construction.

8.5.2   Inside building: Low energy and carbon designs

Buildings consume a significant amount of energy to deliver the conditions for the prescribed use. Designs of the Mechanical, Electrical, and Public Health (MEP) systems in buildings have to safeguard the basic requirements on health and safety. The minimum performances of ventilation, light, fire safety, and public health has to be delivered. Regarding a sustainability perspective, the HVAC, electrical, plumbing, and drainage systems need to operate in the most efficient way to minimise the use of natural resources, in particular energy. These systems account for a significant portion of life-cycle energy or carbon, as high as 80% for commercial offices or 60% for typical residential buildings. A good design that has incorporated operational considerations (on maintenance and risks) is what the industry needs for the next generation of practice.

The challenge of the MEP design is for high-rise buildings, where the issues of coordination of services and energy efficiency are imperative. Nowadays, office buildings can be as high as 600 m tall with GFA as much as 400,000 m² to serve. The flexibility, efficiency, reliability, serviceability, and maintainability are the main criteria for a good design. The advancement in building construction also provides an opportunity to adopt state-of-the-art systems. The followings are some of the technologies and strategies available for current building designs.

8.5.2.1   Under-floor air distribution system

Under-floor air distribution system (UFAD) is more energy efficient as only the lower occupied zones of a space are conditioned, and the higher zones are unconditioned. Energy simulation has demonstrated that it can have the potential saving of 8-10% of the total energy for air-conditioning. It also has the advantage of creating better IAQ due to its nature of displacing polluted air with fresh air at the occupied zone.

The design of UFAD requires due consideration on the placing of terminal unit at the floor. Uncoordinated arrangements may result in constraining the office layout at minimum and inferior thermal conditions at the worst. CFD simulations can predict the performance of the system and determine if local over-heating may be created. The adjusting on the supply and return air temperatures may be necessary if the thermal comfort is a key concern of the space.

8.5.2.2   Radiant cooling

Conventional A/C systems use convection of heat principle to deliver a comfortable space for occupants. This requires fan power to distribute the cool air to an occupied space. An alternative approach of providing thermal comfort is by means of radiation principle of heat. In fact, radiant cooling provides better thermal sensation than convective heat transfer.

As air circulation is not required, radiant cooling can save energy for operating fans. It is estimated that the potential saving can be as high as 5% of A/C energy depending on the cooling load of space. In terms of design, radiance cooling can be coupled with a UFAD system for greater performance and space planning.

8.5.2.3   Daylight and low LPD lighting

Energy for lighting accounts for around 30% of total energy use of a typical office building. The potential of saving is significant when daylight coupled with low Lighting Power Density (LPD) lighting is designed.

The design of daylight has to consider the potential problem of glare and thermal comfort at the perimeter zone. Daylight simulation and thermal study should be conducted to verify if the provision of high-performance glass and shading devices are required to provide quality lighting and thermal environment.

For artificial light design, it is becoming a trend to design with low LPD provisions. The LPD design for offices has not been approached 8 W/2 (from 20 W/m as prescribed by CIBSE Lighting). This is possible due to the improvement on the performance of lighting fixtures. T5 or even LED has much higher efficacy. In addition, it is also a design trend for offices to install task lights that lower background lighting levels form 500 lux to 400 lux or even lower.

8.5.2.4   *Water management*

Water management is essential for sustainable buildings. Many green building rating systems emphasise various categories on evaluating the effectiveness of water use in the buildings. Typical categories are annual water use, monitoring and control systems, irrigation, collection, recycling and reuse, and water treatment systems, and so on. Many systems and technologies are available to enhance water management and it is important to incorporate this important aspect early in the design process. Example 8.7 shows the achievement of water management through the Vanke Centre project in China.

Example 8.7   Vanke centre in Shenzhen, China

Vanke is the most influential property developer in Chinese Real Estate and is also one of the largest in the world. Vanke's vision for its headquarters in Shenzhen is to confront and promote sustainable, healthy, environmental and socially responsive urbanisation. The iconic, elevated and longitudinal form respects local generic context, and Vanke Centre, as seen in Figure E8.7.1, realised as "a horizontal skyscraper over a maximized landscape". The "Sea Scribble" ecological garden embraces and is simultaneously reinterpreted among horizon, spaces nature and individuals within.

Achieved the LEED platinum certification, Vanke Centre utilised cost-effective sustainable design solutions focusing on adaptation, low resource consumption, environmental friendliness, and social harmonious aspects. Various sustainability features are on water management, energy and system, material and recycling, well-being, and social benefits and community development.

For water management, the Vanke Centre adopted a landscape integrated rainwater harvesting system with a capacity of 1,200 m³, grey water recycling that saves water up to 150 m³/day, and waterless devices that reduces the equivalent water usage of up to 51%. Storm water management for speed, quantity and water quality was achieved through the collection and filtration of an artificial lake and the coverage rate of landscaping of more than 90%. To meet water quality standards for the wastewater, a treatment plant with mechanical/natural systems is in place.

Figure E8.7.1   Vanke Centre in Shenzhen, China. © Hufton+Crow

8.5.2.5   *Aquaponics*

An aquaponics system consists of a fish pond and a planting area that could be in the form of a constructed wetland or hydroponics area. The system mimics the characteristics of natural ecology for water recycling and requires less energy for water treatment compared with other technologies. The proposed system will treat and filter fish pond water through the roots of plants before re-circulating the water back to the fish pond or introducing the water for irrigation. Edible fish and vegetables could be farmed in the aquaponics system.

8.5.2.6   *Energy use target*

During the design stage, it is important to set a target for energy saving as compared with the baseline case. Normally, the baseline case is the best practice in according the energy regulations (ASHRAE Standards for LEED and BEC standards for BEAM Plus buildings). Energy saving can be achieved using different passive designs (reducing the demands) and active systems (enhancing the efficiency). The MEP design has to be achieved later using state-of-the-art technology. Therefore, some sustainability practitioners have referred to it as a high-tech solution. It requires the understanding of the performance of individual operations under different conditions, the internal loads, external weather conditions, and more importantly the potential risks of malfunctions.

The practice of green building in the region in the past 20 years has amassed much information on the efficient systems and their potential savings for different uses of buildings under particular climatic conditions. Energy modelling is a useful tool and a critical process for evaluating the effectiveness of design under different climatic and operational conditions. They are critical for decision makers to make informed decisions based on the investment of energy saving systems.

It is now known to practitioners that the portions of energy use from different systems (e.g., HVAC) amount to around 50% of total energy usage and 25% for lighting. It becomes apparent for them to focus on these areas when energy saving is a key component of a design. Advanced technology can be applied step by step on arriving at the target of a design. A review of these different technologies will shed the light on the solution.

8.6   SUMMARY

The practice of sustainable building is still at an early development stage. Standardisation processes are underway to help make our future buildings more adaptive and resilient to climate change or climate extreme. Advancements in technology in various aspects have made our buildings capable of withstanding extreme conditions as a result of climate change. High-performance buildings and innovations are now becoming mainstream in the market for wider adoptions. These movements are driving practitioners to build even better buildings.

Chapter 9
De-carbonisation

9.1 INTRODUCTION

Design professionals around the world have been searching for solutions to de-couple our living habits from carbon emissions. Since buildings contribute a significant portion of the total amount of carbon emissions in cities, many design strategies have been developed to tackle the problem. In the big cities of Asia, the concept of Zero Carbon Building (ZCB) or Positive Energy Buildings is popular. Demonstration projects have been implemented to test the viability of new technologies and acceptance from the occupants.

In this chapter, building energy performance is discussed, firstly to gain an understanding and benchmark the existing building systems. Such understanding will form a basis for identifying the areas that need to be focused on and for formulating the appropriate strategies and technologies that can be adopted for effective improvement. The process leading to a zero carbon society is a long one and experimental in most cases. It is imperative that designers keep in mind the gap between the ultimate vision and reality, as well as the importance of a step-by-step approach. Cases of ZCB will be used to elaborate further on the design process and on the performance of buildings during operation. Discussion will particularly be focused on the constraints of ZCB in the urban context of cities, by means of renewable energy within a dense urban environment.

Although achieving zero carbon at the individual building level is essential, it may not be possible in the wider context. Therefore, district energy with connection to a micro-grid network could improve the overall performance for a district or city. By formulating centralising solutions in order to improve efficiency from district energy and decentralising solutions to improve flexibility and reduce loss from distributed energy generation, de-carbonisation at the city level is made possible.

Building Sustainability in East Asia: Policy, Design, and People, First Edition. Vincent S Cheng and Jimmy C Tong.
© 2017 John Wiley & Sons Ltd. Published 2017 by John Wiley & Sons Ltd.

9.2   BUILDING ENERGY PERFORMANCE

In recent years, conducting energy audits on buildings has become mandatory in many Asia countries and they are required to make the results available to the public. The data shows how existing buildings of different ages are performing, and how they compare to new buildings with the latest technologies and updated codes of compliance. The benchmarking of performance is helpful in identifying which building components should be the focus for improvement.

One of the indicators for building performance is energy use intensity, which is defined as annual electricity consumption per unit area, in terms of $kWh/m^2/yr$. While it is helpful because it is easy to understand, it is challenging to gather the data for a fair comparison. It is worth mentioning that the building energy performance is a composited result from the design and operation of the building. Some of the strategies for improving performance have been described in the previous chapter. For meaningful benchmarking, it is important to recognise that the different energy consumption patterns/profiles from the various types of building are not identical. Therefore, the development of a consumption model for each building type is necessary. The main categories are residential, commercial, and industrial, and it is also important to create further subcategories for the benchmarking data to be productive. Not only the types of building require breaking down, but it is also important to keep track of the types of energy use. Example 9.1 will illustrate the application of such a process. This provides information for the existing performance and a reality check for achieving zero energy objectives.

Back to using carbon emission as a measure, which can be converted from the building energy intensity, Figure 9.1 shows an estimation by a study from the

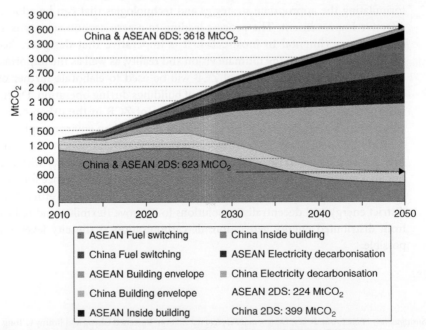

Figure 9.1   Carbon Emission Projection from Buildings in China and ASEAN. © Arup

International Energy Agency (IEA). The figure covers the carbon emissions from the building sector in China and countries from the Association of Southeast Asian Nations (ASEAN) from 2010 to 2050. The projection of 3,618 $MtCO_2$ results from a scenario commonly referred to as the six degree scenario (6DS). If proper mitigation strategies are applied, it would be possible to attain the two degree scenario (2DS), where carbon emissions by 2050 would be only around 47% of the emissions in 2010. This illustrates how important building performance can contribute to carbon reduction (Example 9.1).

Example 9.1   Building electricity use intensity in Hong Kong, China

Research was conducted to benchmark the building performance for 2005. For many years, buildings in Hong Kong have been consuming about 90% of the electricity and contributing around 67% of the carbon emissions. As a result of Hong Kong's economic restructuring and relocation of traditional manufacturing activities to the Mainland, the demand for private industrial buildings is decreasing. Therefore, the research was focused on the residential and commercial sectors. Both sectors were then further divided into several building subcategories,[115,116] with categorisation mainly based on the standards of the EMSD when publishing their annual energy end-use data. In the end-use data, the commercial subcategory consisted of retail, offices, restaurants and hotels, health, and storage buildings. It was calculated that the other commercial segments in the end-use data covered about 40% of the energy consumption in Hong Kong.

The various divisions allowed the major electricity end-uses and potentials for electricity saving in each building subcategory to be identified. The consumption model of each building subcategory was constructed using a computer simulation, which was based on information collected from past projects and the daily routines in each building subcategory. To ensure the accuracy of the models, a calibration process using real-measured data was conducted. In order to assist the collection of real-measured data, a set of questionnaires was delivered to the relevant departments, including CLP Power, Hong Kong Electric, Towngas, the EMSD, Rating and Valuation Department, and the Housing Authority. Moreover, the data can also be extracted from research papers, which were written based on statistical studies conducted by scientists. It was not a straightforward process as this data came from various sources and it was difficult to normalise the common definitions among them.

Electricity consumption models for ten different building segments have been constructed using computer simulations, and are summarised below in Table E9.1.1. The table indicates that the consumption intensity in restaurants is the highest among all the building subcategories, followed by retail, hospitals and clinics, and finally office buildings. The high numbers indicate that the potential for energy savings in commercial buildings must be huge. For the residential sector, the variation in electricity use intensities are revealed, in which the private housing subcategory has the highest intensity.

The electricity use intensity for each building type was then compared to data obtained from various sources, including research papers and EMSD Energy Consumption Indicators, as well as Online Benchmarking Tools.[121] Table E9.1.1 indicates that restaurants consume the largest electricity use intensity among all building types, followed by retail buildings. Restaurants consume much higher energy due to their cooking function. Based on the EMSD energy end-use data, cooking accounts for 50% of the energy demand in restaurants.

Moreover, it can be seen that retail buildings consume more energy compared to office buildings, as most retail buildings operate longer hours than office buildings in a single year. Furthermore, the energy demand for lighting end-use in retail buildings is significantly larger than the demand in office buildings. On the other hand, although hotels have longer operational hours than offices, it is

Table E9.1.1 Summary of Electricity Consumption Models

Building Sector	Building Subcategory	Baseline Electricity Use Intensity in 2005 (kWh/m²/year)	For Comparison Purposes Data Collected from Various Resources (kWh/m²/year)
Residential	Public Housing	115	82–155[117,118]
	Private Housing	135	
	Subsidised Flats	100	
	Other Housing	85	
Commercial	Retail	660	639[116]
	Office	415	18–504[116,119]
	Hotel	410	195–683[116,120]
	Restaurant	1,350	411–1,839[116]
	Hospital and Clinic	425	105–475[116]
	Storage	355	356[116]

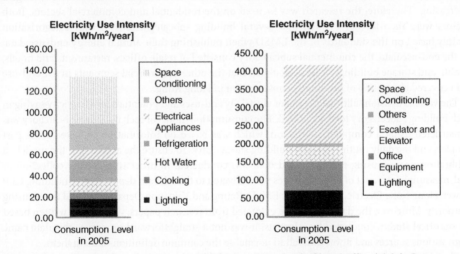

Figure E9.1.1 Electricity Consumption Model of Private Housing (Left) and Office (Right). © Arup

found that the electricity use intensity in offices is slightly higher than in hotels. The study reveals that electricity is used in hotels to meet 78% of its energy demand. In contrast, electricity is used to meet almost all of the energy demands in offices. Thus, as expected, the energy use intensity of hotel buildings is actually larger than that of office buildings.

The figures below present the electricity consumption for different end-uses in private housing and office buildings. As for office buildings, Figure E9.1.1 shows that space conditioning systems, office equipment, and lighting systems account for about 52%, 19%, and 17% respectively, of consumption. By noticing the profiles of energy usage, it is helpful to look for strategies and technologies to target the correct category for energy consumption reduction.

Furthermore, the net electricity demand in Hong Kong for 2005 is illustrated in the following figure, which presents the electricity use intensity (EUI) and floor area of each building subcategory. Thus, the electricity consumption of each building subcategory is proportional to the area of the shaded box of the corresponding subcategory, as shown in Figure E9.1.2. Again, similar to knowing the profile of each building type and even individual building performance, with the information of the weight/impact for each type of building by means of total floor area, proper follow up actions from policy to individual choices can be determined accordingly.

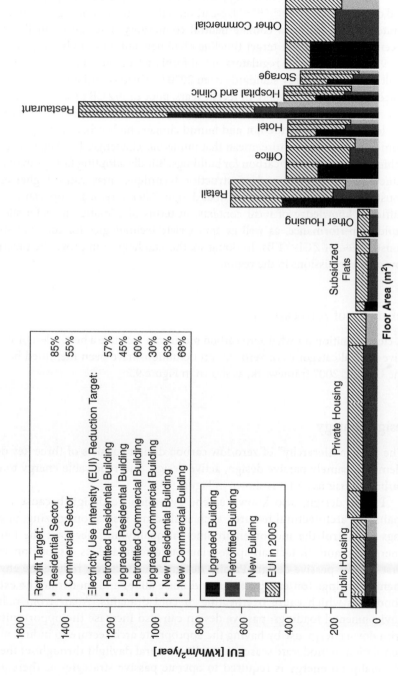

Figure E9.1.2 Retrofitting and EUI Reduction Targets for Different Types of Buildings in 2030. © Arup

9.3   LOW/ZERO CARBON DESIGN

To improving the current building performance levels, Zero Energy Building (ZEB) and Zero Carbon Building (ZCB) are becoming a globally popular concept and design trend.[122,123,124,125,126,127] De-carbonisation is the ultimate goal of one of the strategies for creating a climate neutral community. The European Parliament recently announced the target timeline of all new buildings to be zero energy by 2019.[128] In 2007, energy regulators in California set a target of all new homes to be built to net zero energy standards from 2020.[129] Various initiatives/policies are in place to accelerate the trend. In recent years, pilot ZEB/ZCB projects were built as key initiatives for governments in Europe, and it is now happening all across the world. In Asia, due to the hot and humid climate, high urban density, as well as human activity and culture, mean that numerous challenges lie ahead in order to achieve net zero energy/carbon for buildings. Blindly adopting European or North American architecture and construction techniques may cause higher energy consumption rates than conventional design. Therefore, it is important to have sufficient knowledge of local contexts, in terms of climatic characteristics and building performance, as well as appropriate technologies for the analysis and construction of ZCB/ZEBs. In doing so, this can help to improve the practice of low carbon emissions in the region.

9.3.1   Definition of zero carbon

A clear definition for what zero carbon means can facilitate a better design strategy. Five general categories of zero carbon definitions have been identified based on the UKGBC 2007 framework, as shown in Figure 9.2.

9.3.2   Design strategy

The "golden hierarchy" of zero/low carbon design consists of three key design elements, namely passive design, active design, and renewable energy to offset carbon emissions and it is shown in Figure 9.3.

Passive design, also known as energy-avoidance design because it relies mainly on architectural features, such as shading, insulation, glazing, or openings to control the indoor environmental quality and decrease the building energy demand. In fact, an appropriate build form and its disposition are the first step for passive design. Passive design can effectively reduce the environmental loadings (on energy, solar, and noise alike) and moderate the extreme conditions which cause overheating or cold draughts. Apart from reducing environmental loadings, passive design can also increase the opportunities to trim down energy use by having the appropriate architecture to induce natural ventilation in moderate seasons as well as natural daylight throughout the year. Normally, no energy is required to operate passive strategies as there are no

Hierachy of Low and Zero Carbon Building Definitions

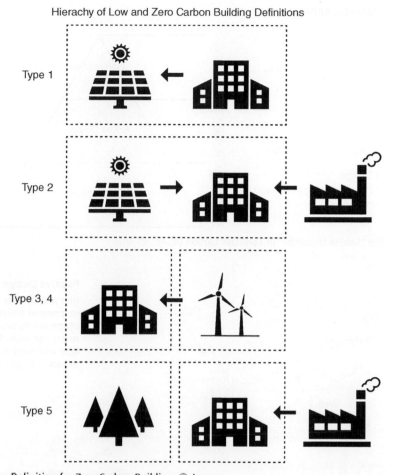

Type 1

Type 2

Type 3, 4

Type 5

Figure 9.2 Definition for Zero Carbon Building. © Arup

mechanical parts, and therefore the life-cycle energy is low. Example of passive design is illustrated in Figure 9.4.

Designers must have good knowledge of local climate characteristics, and must create responsive architecture that can balance the needs of indoor activities and the vision of low/zero energy use. Successful passive design can be integrated into the overall architectural design and enhance the aesthetics of the building.

Active design, also known as energy-efficient design because mechanical and electrical systems are incorporated to deliver the services needed to meet the design requirements of the function of the building. For some building types, for example, office buildings located in sub-tropical regions, passive design alone cannot make them zero carbon. This is due to high-energy demands in local weather conditions and natural activity, such as air-conditioning on hot summer days and

ZERO CARBON APPROACH

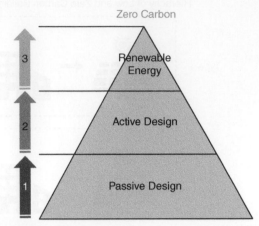

Zero Carbon

Renewable Energy

Active Design

Passive Design

Renewable Energy
Reduce Carbon Emission using renewable energy systems

Active Design
Enhance Building Energy Efficiency through utilizing energy-efficient systems and equipment control

Passive Design
Reduce Building Energy Demand through natural ventilation, daylighting, solor shading, etc.

Figure 9.3   The "Golden Hierarchy" of Zero/Low Carbon Design. © Arup

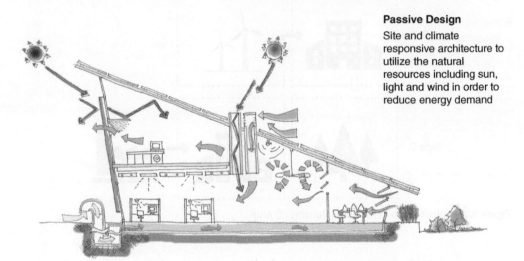

Passive Design
Site and climate responsive architecture to utilize the natural resources including sun, light and wind in order to reduce energy demand

Figure 9.4   Passive Design. © Arup

lighting during the night-time. Energy is still required and appropriate building systems are critical. Example of active design is shown in Figure 9.5. Efficiency is the key to the design. Systems can be efficient when the services are delivered on-demand, on the spot, and recycle where appropriate.

Renewable energy to offset carbon emissions is the third and final element of zero carbon design by using said renewable energy to balance out the fossil energy used to power the active systems. Subject to availability this includes solar, wind, bio-fuel, hydro, marine, and geothermal energy.

Generally speaking, zero carbon buildings should have included the idea of renewable energy application at the early design stage. Build form and architectural shape affects the natural resources available such as solar and wind energy.

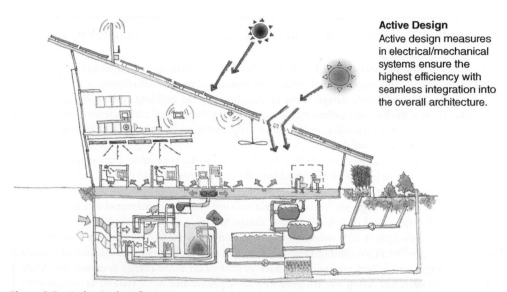

Active Design
Active design measures in electrical/mechanical systems ensure the highest efficiency with seamless integration into the overall architecture.

Figure 9.5   Active Design. © Arup

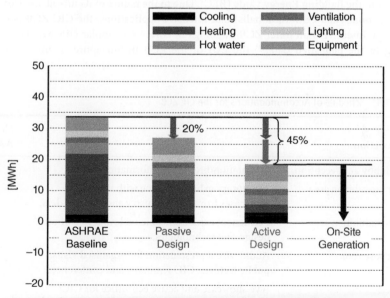

Figure 9.6   Example of a Step-By-Step Zero Carbon/Energy Design Approach. © Arup

Analysis is required to determine the best building location and form to make the most out of the natural resources. Figure 9.6 shows how each of the steps can contribute to the energy or carbon saving for the building project, and Example 9.2 is a good example of a zero carbon building that was brought to reality using this design approach.

Example 9.2   Construction Industry Council ZCB in Hong Kong, China

The Construction Industry Council ZCB (CIC ZCB) is a net zero carbon building (Table E9.2.1 and Figure E9.2.1) that was designed around the typical hot and humid climate of sub-tropical Hong Kong. The sustainable design of the architecture and the building systems cover various considerations of the life-cycle carbon emissions, including the inherent carbon from construction materials, emissions associated with the construction process, and the 50-year operation and decommissioning of the building.[130]

The total life-cycle carbon emissions were offset by on-site renewable energy generated by photovoltaics (PV) and bio-diesel combined with the cooling, heating and power (CCHP) system. A number of sustainable strategies on passive architecture were used to optimise the design of CIC ZCB. They included the high-performance façade with low overall thermal transfer value (OTTV), effective air tightness, and optimised window design that allowed for the application of natural ventilation and daylight. A 20% reduction in energy demand was achieved through these passive designs. Energy-efficient air-conditioning (A/C) systems using desiccant dehumidification, under-floor air supply, and radiant cooling also helped to achieve ultra-low energy consumption. More than 45% amalgamated saving was achieved with these passive and active systems as compared to the existing local building energy codes.

For this case study, an energy model was built and run using these values; the predicted energy use intensity (EUI) of CIC ZCB is 86 kWh/m², or 45% less than the compliant baseline building from the Building Energy Code (BEC). Due to the nature of its mixed-use (for offices, conference rooms, and exhibition halls) and intensive applications, the CIC ZCB is relatively more energy-intensive than other ZCB buildings in cities with similar climates. For example, according to the published data,[131] the EUI of the ZCB in Singapore is only 46 kWh/m².

Table E9.2.1   Schedule of Accommodations for the CIC ZCB

Accommodation		Net Floor Area (m²)
Entrance lobby and reception	Orientation/break-out/information;	100
Temporary exhibition area	Temporary exhibition zone with renewal showcases from local industry and stakeholders;	150
Permanent exhibition area	Permanent exhibition zone on low/zero carbon design and technologies;	490
Multi-purpose room	Audio-visual presentation for organised visits/ public lectures, CIC seminars, and conferences;	260
Eco-office 1	Live showcase and active eco-office for CIC;	230
Eco-office 2	Live showcase and active eco-office;	120
Eco-home + display gallery	Demonstration of low/zero carbon home design, features, and engagement for low-carbon living;	150
Souvenir shop	Souvenirs/eco-product retail space;	10
Eco-café	Ancillary catering facilities of the ZCB, with an eco-theme for sustainable food;	10
Total net floor area		1,520

Figure E9.2.1   CIC ZCB in Hong Kong. © Arup

The function and operation schedule of the premises could lead to a large difference in annual energy use.

The building and landscape energy consumption is greatly reduced by applying the passive and active design measures for energy efficiency. For example:

- Building: 116 MWh/yr electricity consumption
- Surrounding landscape: 15 MWh/yr electricity consumption

To achieve net zero carbon, these energy demands were met through use of renewable sources. An additional 99 MWh/yr was exported to the grid to offset the embodied energy in major building materials.

With limited roof space for mounting the PVs as well as other site constraints, it was concluded during an early feasibility study that power from PVs alone would not be sufficient to meet building loads. The CIC ZCB is surrounded by other buildings. There is one high-rise office tower on its south side which completely overshadows CIC ZCB during the winter solstice. The expected power output per square metre of panels is approximately 85 kWh/m². The panels will also require approximately 50% mounting and servicing space. The PV cells themselves are monocrystalline and polycrystalline systems. Complete coverage of the building footprint with 1,015 m² PV panels will produce only 87 MWh/yr of electricity – that is, insufficient for demand.

A bio-diesel CCHP system is a suitable additional renewable energy source as the fuel is locally supplied and is produced from waste cooking oil.[132] A CCHP system is suitable as the thermal energy produced can be used to drive an adsorption chilling system which will provide cooling to the building, as a result, it cuts down 85% electrical energy used for the chillers. Figure E9.2.2 summarises the energy balance with the application of bio-diesel CCHP and PV systems.

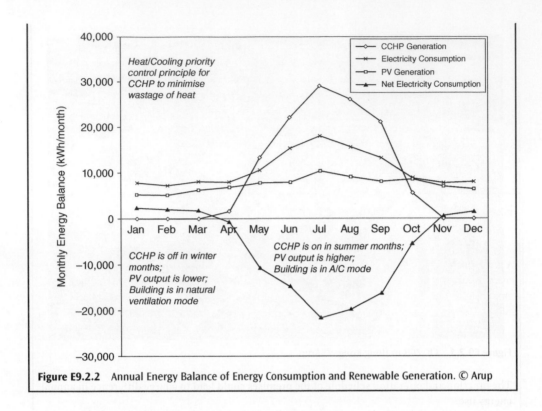

Figure E9.2.2   Annual Energy Balance of Energy Consumption and Renewable Generation. © Arup

9.4   RENEWABLE ENERGY FOR URBAN DEVELOPMENTS AND BUILDINGS

Many strategies for both passive and active designs have been covered in Chapter 8. Given the global trend to move our reliance from fossil fuels to renewable energy sources, it is important to identify the effective approaches and steps to promote a wider use of renewable energy:

- At a policy level - regional resource assessments, and regulatory requirements and incentives.
- At a project level - feasibility studies, technical evaluations for specific sites, financial or social responsibility justification, budgeting, detailed design, sourcing, and implementation.

In the urban environment, renewable energy can be harvested at the district/utility level or the cluster/individual building level. At the utility level, large-scale onshore and offshore wind farms can be found all over the world. Large-scale solar farms, bio-energy and geothermal power plants, and various sizes of hydro power generators are very common depending on the favourability of the geographical location. Only large-scale marine/ocean power generation is still in an early development stage. Large-scale renewable development plays an important role in the de-carbonisation process as they are more efficient and cost effective. However, small-scale renewable generation will have much wider application since it is a distributed source that is closer to the demand.

In the urban built environment, the density of buildings causes significant constraints on installation. Successful application of renewable energy, especially for building-integrated systems, will rely heavily on building physics. The availability of most renewable energy is highly dependent on local conditions, about which building physics analysis can provide useful information. Example 9.3 is an example as to how this is relevant for PV in Hong Kong. For instance, when it comes to wind, hydro, and marine energy, computational fluid dynamics is the

---

Example 9.3   PV performance in the dense urban environment of Hong Kong, China

A greater use of renewable energy can alleviate urbanisation problems. There are various renewable energy sources at different stages of development. Among them, solar, wind, and bio-energy are more mature; but each of them has its own pros and cons. Given the density of high-rise buildings, building-integrated renewable solutions are required. Whereas utility-scale renewable power plants require land space that the city does not have, building-integrated renewable power generation can directly supply electricity to meet the loading demand, thus minimizing transmission losses. Among mature renewable technologies, solar PV has the longest history as a building-integrated solution. Given the complexity of cities with high-rise buildings, each constructed for various purposes and of different shapes over a wide period of time, PV application faces many challenges with new building developments as well as existing building retrofitting. An understanding of this urban effect on PV performance is important when estimating the overall potential for this type of technology.

In 2004, the Hong Kong Special Administrative Region Government conducted a study called "A Survey on Implementation of Photovoltaic Systems in Hong Kong". The purpose of the study was to determine the allowable space, technical issues, costs, possible site-specific constraints, and barriers to the installation of both Building Integrated Photovoltaic (BIPV) and non-BIPV systems in Hong Kong and to evaluate the electricity generation potential for territory-wide implementation using PV technology.

To evaluate solar energy potential in the urban environment, not only are local solar radiation and cell efficiency major factors, but so too are the individual and collective attributes of buildings and open spaces. These attributes are not only important for the evaluation of individual performance, but the interactive effect of these attributes among buildings and open spaces will affect the overall solar energy potential in an urban situation. For instance, the attributes are orientation, height, footprint area, and building age, as well as the shadowing effect from the interaction of these attributes.

From the sample locations of buildings and open spaces, applicable areas for possible PV installation on rooftops, façades, and in brown area sites were identified. In these representative locations, the mean with statistical uncertainty values in the applicable areas were calculated and expressed as a percentage of the footprint area for the locations. The values were then applied to the collected footprint areas of shortlisted locations for each category. It is worth noticing that about 5.4% applicable rooftop space of the overall footprint area and about 4.1% applicable space for the overall façade area from the list are considered feasible for PV installation. This is taking into account the age, shadowing effect, and minimal cost-effectiveness of the area. As for the open space, it was found that more applicable area was available than the estimated 2% to 3.3% from the survey. Together with solar radiation and cell efficiency, with respect to respective orientation, the annual PV generation potential for existing buildings and open spaces is 71.4 GWh and 56.6 GWh, respectively for BIPV and non-BIPV.

tool used to identify the zones with a higher air and water flow to generate more energy. For solar power, the path of the sun and the shading effect from the surrounding environment has a significant influence on performance. With the application of bio-energy, determining the suitability for integrating the generator into the building requires a lot of design consideration. When designing chimneys and flues it is important to comply with the relevant regulatory and environmental codes, as well as properly integrate the system with other heating and cooling systems which in itself can be a complicated process. When it comes to geothermal energy, the size of piping and the integration of the pump and thermal cycle systems need a basic knowledge of the fundamentals of building physics. As this section begins at the building level, the following renewable applications will focus on building-integrated/small-scale solutions.

9.4.1   Solar energy

Solar energy is one of the most popular renewable sources because it is considered abundant and accessible. At the building level, both BIPV and non-BIPV can be grid connected or a stand-alone system. Because the possible solar energy harvest throughout the building is limited due to environment constraints, the generated power is only used to meet local demand; when excess power is available, the power can transfer to the grid for others to use. When solar panels or arrays are designed to be incorporated onto rooftops and façades of buildings, they are considered as being BIPV; when systems use areas other than building structures, such as open spaces, hill slopes, roadside noise barriers, and bus shelters, they are considered non-BIPV.[133] In order to be completely operational, solar systems also include solar PV cells, but the overall system also consists of inverters, energy storage, and monitoring and metering devices (Example 9.4).

Example 9.4   PV design for the Hanwha Headquarters building retrofit in Seoul, South Korea

Arup have been commissioned by UN Studios to develop a new façade system for the Hanwha Headquarters building (Figure E9.4.1) in Seoul, South Korea. The original building was built in the 1980s. With ever advancing façade technologies, the original façade system has become inefficient by today's standards. Further to designing an iconic new façade for the building, UN Studios also intends to integrate Hanwha's PV panels from its core business into a complex façade system to display a vibrant pattern on the exterior. Arup has been appointed to develop technical details for the façade system and to carry out sustainability studies that provide a foundation for further development.

The Hanwha Headquarters is a 29-storey building with an approximate total Gross Floor Area of 57,683 m². It is located at 1 Janggyo-dong, Jung-gu, Seoul, South Korea. The building is surrounded by commercial, financial, and government buildings, other corporate headquarters, and cultural assets. Using PV panels is an excellent way of generating renewable electricity from solar energy. The Hanwha Headquarters has to be able to generate 3% of its energy consumption from renewable sources as a key performance indicator of the corporate strategy.

Figure E9.4.1   Hanwha Headquarters in Seoul, South Korea. © UNStudio

This study aims to provide a foundation for developing an optimised PV system. The model of PV panels was chosen by Hanwha and some important information is listed below:

- The PV panels could be mounted on the sloping or vertical surfaces of the atypical façade modules. However, it was found that mounting the PV panels vertically would result in a 30% drop in capacity. It was, therefore, recommended that the PV panels be mounted on the sloping surface to increase system capacity.
- The panels will be situated on the sloping elements of the atypical facade to maximise the capacity of the system and thus to act as a visual representation of Hanwha's technology and approach.

Based on the Korean guidelines, PV panels should receive more than five hours of daylight. As the south façade might be shaded by the surrounding buildings, a first design approach will be to use a daylight access analysis from the building physics to determine possible installation locations. With these basic requirements in mind, a percentage shading analysis was carried out to determine the possible manoeuvring of the PV panels to optimise their performance.

A software program called Ecotect was selected to perform a simulation for both the "daylight access analysis" and "PV % shading analysis". The winter solstice was used as a design case for the simulation because daylight duration would be the shortest and shading from the surrounding buildings would be the heaviest. This will ensure that the PV panels comply with the local regulation at all other times of the year as well.

The cumulative hours of daylight that fall on each panel is shown in the "Daylight Access Analysis". The panels do not comply with the Korean guidelines and may have to be removed. From the results of the "PV % shading analysis", the PVs were categorised as (1) may have to be removed and (2) should be relocated because they underperform.

An alternative performance-based approach was also used to determine the location of PV panels on the facade. In this methodology, the typical case referred to was found in the Chartered Institution of Building Services Engineers (CIBSE)'s publication, "Capturing solar energy (2009)". A benchmark of 800 kWh/m^2 of solar irradiation per annum was set to determine if a location was suitable for installing PVs.

Using this approach, many of the PV panels at the upper and lower part of the South facade should be removed, when compared with the first approach. However, this approach allows some PV panels to be installed on the East facade. However, reducing the number of PV panels will reduce the power generation capacity of the facade. To supplement the energy production of the PVs, the roof levels should also be utilised. It is proposed that an estimated 364 PV panels be installed.

In conclusion, the first approach adheres better to the architect's design intent by allowing more PV panels to outline the profile of the dynamic "eye" zones on the façade. While the second approach guarantees the efficiency of the PV panels and may be more cost effective from a payback and power generation point of view.

9.4.2   Wind energy

In small-scale wind power generation, although there is no standard definition, it is typical to have a capacity under 100 kW and under 1000 kg in weight.[134] Both horizontal and vertical turbines are commonly used in a building-integrated system. Besides the space constraint, the concern for available wind direction plays an important role in determining which type of turbine is more advantageous. In an urban environment, not only is wind power found in building applications, but also in standalone applications such as lamp posts and communication towers where wind power provides energy directly for lighting and cellular phone equipment.

9.4.3   Bioenergy

Direct disposal of organic waste in cities increases the landfill load with a heavy demand on the logistics required to transport the waste to the site. This indirectly produces carbon emissions. To mitigate the environmental impact in dealing with such a huge amount of waste, it is important to convert this waste into a useful form of energy. Biomass has been used to generate electricity for over 100 years and to generate heat for many thousands of years. Biomass Combined Heat and Power (CHP) technologies are now available in capacities ranging from a few kilowatts of heat and power to those able to generate megawatts of heat and power.

Due to the ability of biomass CHP technologies to supply de-carbonised heat and electricity, with fewer of the limitations of other renewable energy technologies used in the built environment, biomass CHP is increasingly of more interest. There are various types of biomass CHP technologies, including gasification, the organic Rankine cycle turbine, the indirect air turbine, and the steam turbine.

The gasification system turns the stored biomass into a useable fuel known as synthesis gas or biogas through a process called anaerobic digestion.

The generated electricity can be used for building systems and equipment onsite, for example, supplying electricity to the chiller plant operation, and any excess can be fed into the public grid. At the same time the heat, a by-product of the process, can be used to dry the biomass and may be used to speed up the anaerobic digestion process and thus, the biogas generation rate by maintaining the gasifier at a certain temperature. Other forms of heat can be supplied for building uses, such as hot water provision to catering facilities. In fact, biomass technology is well-developed in many countries, and includes a variety of applications for example on farms and rural domestic properties, and in commercial buildings and facilities.

9.4.4 Hydropower

Hydroelectricity has been widely considered as a green and low-cost approach to generating electricity with the added bonus that it makes use of the unused water head in potable water pipelines. This makes it a promising and sustainable energy solution without affecting the water supply to buildings. Water moving along pipelines carries substantial energy, and Hong Kong, for example, has a robust and sophisticated water infrastructure, which represents a wealth of energy that can be tapped into to generate power in a sustainable way. In addition, Hong Kong has moderate per capita energy consumption, and the energy-saving potential in high-rise buildings in urban areas is significant. To reduce the reliance on fossil fuel-based power, hydroelectric power from buildings has a great deal of potential to grow, including wider applications such as grey water or used water (Example 9.5).

Example 9.5   Inline-hydro power turbines in high-rise buildings in Hong Kong, China

A research project on the first in-building hydro power turbine was commissioned by the Sino Group, Arup, and the Hong Kong Polytechnic University. The in-building hydro turbine is designed to reclaim the unused water head in pipelines. A prototype at Olympian City 2 (Figure E9.5.1) has demonstrated encouraging preliminary results that could generate hydroelectricity in a green and low-cost way while reducing carbon emissions.

With an output power of 100 W, the prototype is the second phase of hydro power generators which has seen a six-fold leap in power generation from the first phase. Hydroelectricity from water in pipelines is generated when water current passes through a vertical-axis turbine developed by the Hong Kong Polytechnic University and the Water Supplies Department. The electricity generated is then stored in a battery system and is used to power a typical lift lobby's lighting system. The lighting system in the lift lobby has 15 LED lights each requiring 6.5 W of power; the in-building hydro turbine could save around 700 kg of carbon emissions annually, which is equivalent to the offset of 30 trees.

Figure E9.5.1   In-building Hydro Turbine Installed in Olympian City 2, Hong Kong. © Arup

9.4.5   Marine/ocean energy

Cities along the coast lines and rivers have access to a constant and reliable source of energy from the movement of the water. The development of marine/ocean energy as a renewable energy source is on the rise globally and its application can be broadened to sea current channels, river channels, the seabed level, reservoirs, and dams. It is also important to establish an understanding of sea movement behaviour, thereby selecting the most optimal location, and accurately estimating power generation (Example 9.6).

9.4.6   Geothermal energy

The ground at sufficient depth, with its constant temperature throughout the year, is a good climate modifier to moderate the extreme conditions experienced during summer and winter time. It is also a good form of renewable energy. In order to achieve energy saving and environmental protection objectives, a closed-type ground source heat pump system can provide cooling in the summer and heating during the winter suitable, for example, to a Beijing-like climate. A ground source heat pump (GSHP) is regarded as one type of renewable energy, which is encouraged under the LEED framework.

A GSHP is used to release heat into the ground for cooling, and to absorb heat from the ground when heating. In other words, during the summer, a GSHP uses the earth as a heat sink while during the winter it uses the earth as heat source. By using the moderate temperatures in the ground, the GSHP increases operational efficiency and reduces the operational costs for the overall heating and cooling systems. GSHP systems reach fairly high efficiencies (coefficient of performance (COP) ranges from 3-6) on the coldest winter nights, compared to a COP of

Example 9.6   Micro marine turbines in low current coastal urban areas

The Arup Building Sustainability Team has collaborated with the Sino Group, City University of Hong Kong, and Inha University of Korea to develop and test a micro marine turbine in Hong Kong. The idea behind the project is to investigate the feasibility of this type of technology in generating electricity for Hong Kong, a city with low tidal conditions.

The turbine was installed on July 2015 on a site owned by the Sino Group (Figure E9.6.1) for actual real-life testing. The turbine was designed for a velocity of around 1 m/s, with the energy generation enough to power a lighting system designed for demonstration purposes. While the turbine was developed by Inha University, Arup has created a Computational Fluid Dynamics (CFD) study to provide a resource assessment for the tidal movement and seawater current velocity of the site. City University and Inha University conducted on-site measurements prior to installation of the turbine, and the data collected was used to calibrate and validate the CFD model.

Figure E9.6.1   Micro Marine Turbine Prototype Demonstration in Hong Kong Gold Coast. © Sino

Table 9.1   GSHP Operation under Typical Conditions

Average soil temperature (within 150 m depth):	14.65 °C
Average soil thermal conductivity factor	2.25 W/mK
Heat rejection capacity in summer (inlet water temperature = 35 °C):	71.64 W/m
Heat absorption capacity in winter (inlet water temperature = 5 °C):	36.07 W/m

1.5–2.5 for air-source heat pumps on cool days. GSHP systems are among the most energy efficient technologies for providing Heating, Ventilation, and Air-Conditioning (HVAC) and water heating. Table 9.1 shows a typical operation of a GSHP and Example 9.7 is a case study for an installation in Beijing.

Example 9.7   BMW green 5S showroom in Beijing, China

A single U-tube will be used as the geothermal heat exchanger (Figure E9.7.1). The size and depth of the hole is 150 mm and 100 m respectively. The efficiency of the ground source heat pump depends on the stability of soil temperature and the heat rejection/absorption contact area. Based on the survey report, the summer heat efficiency and winter heat efficiency of the soil were 71.6 W/m and 36.1 W/m respectively which is suitable for the ground source heat pump system. The estimated ground area for the closed-type ground source heat pump of the kind used in the showroom is about 8,075 m².

The difference between building cooling and heating is about 10% (Table E9.7.1), which can be seen as fairly balanced.

Figure E9.7.1   GSHP System Installed in BMW Green 5S Showroom. © Arup

Table E9.7.1   Operational Data for GSHP in BMW Green 5S Showroom

	Annual total load (kWh)	Heat rejection/ extraction coefficient	Heat rejection/ extraction rate (kWh)
Cooling	1,400,721	1.2	1,680,865
Heating	1,637,248	0.9	1,473,523

9.5   DISTRICT-WIDE DE-CARBONISATION

Although de-carbonisation of a single building is feasible using current technology, de-carbonising an entire district is a whole other problem. It involves many non-technical hurdles, such as urban planning, infrastructure easement, and regulations. It is also prohibitively difficult to de-carbonise an existing community. The planning of new towns provides an opportunity to realise the zero carbon vision on the wider scale.

9.5.1   Micro-energy grid

Energy supply in a conventional district is a unidirectional in nature. Generally utilities provide a service to the building/user with minimal feedback on the nature of consumption. This can lead to inefficient use of resources as consumption cannot be analysed or controlled.

A micro-energy grid (MEG) is a new approach which supplies major utility sources with power combined with a feedback loop and real-time controls to reduce both peak demand and energy consumption.[135,136] Electrical, heating, cooling, and gas district networks are centrally controlled and monitored via the Total Operation Centre (TOC). This facility acts as the nerve centre of the network, by collecting, analysing, and controlling real-time demand data in order to provide energy in the most efficient way possible (Figure 9.7).

By using this smart grid approach each utility can be controlled on multiple levels, for example:

- Micro/room scale - lighting and air conditioning can be controlled remotely to switch off when a space is not in use. This can be controlled either automatically or manually through the TOC.
- Macro/building scale - cooling and heating loads of the whole building can be controlled to adapt to the real-time weather conditions (weather compensation) as supplied by the centralised weather station.
- Major/district scale - generation and supply of district energy is controlled to react to the most advantageous conditions. For example, if a cheaper electricity tariff is available at night the system will generate heating, cooling, and

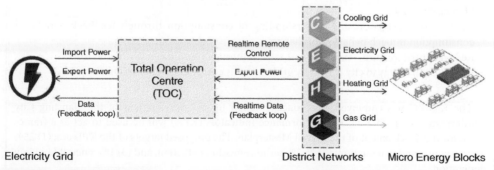

Figure 9.7   MEG Concept. © Arup

electricity then store this energy for use during the day when the peak demand is high, thus reducing peak demand and total energy consumption.

With higher efficiency, reliability, and security, MEGs can meet urban energy demands by connecting on-site or distributed energy generation from individual buildings or communities to the district energy systems. A MEG consists of renewable power generation, power storage, and power demands from commercial, residential, and industrial buildings; it can also include monitoring and control systems and can connect to the main grid structure for backup. A MEG facilitates energy balancing between the loads within the network and between local supply and demand.

To attain higher security from cyber and physical attacks on a local community, special measures can be incorporated into the MEG. It is easier to balance energy use within a community of manageable size rather than doing it on an entire city level. Many MEG examples have been in operation and are well documented.[137] One such example is illustrated in Example 9.8. In addition to the technological aspects of the MEG, regulations are also required to foster the wider application of MEGs as more robust systems.

Example 9.8   K-MEG in Hansung City, Qingdao, China

A key concept of the Korea Micro-Energy Grid (K-MEG) is the application of the energy cascade principle. In many instances, energy is wasted through the energy generating process. For example, around 70% of the energy contained in fuel is wasted as heat loss in conventional power generation. This "waste" heat can be recovered and used for generation of further resources (heating and cooling) which increases the overall efficiency of the process and reduces the total energy consumption and carbon emission levels.

Through the adoption of the K-MEG concept many benefits have been realised, including:

- Demand response serves to reduce peak demand on local utilities and carbon emissions
- Real-time monitoring and response to manage demand and energy
- Incorporating the energy cascade principle to increase efficiency and reduce carbon emissions
- Open protocol information network to allow maximum connectivity and flexibility
- Resilience/reliability of supply to ensure no utility downtime – essential for the medical nature of Hansung City
- Greater user awareness and understanding of consumption through feedback – reduced consumption through behavioural changes
- Reduced infrastructure – time and cost savings through coordinated district networks
- Export clean energy to city electricity grid – surplus energy exported back to city grid

The K-MEG was employed as a concept design for the development of the Hansung City Masterplan (Figure E9.8.1). Key Performance Indicators (KPIs) will be used to act as a framework for the development of the overall Masterplan. The proposed targets of the KPIs are: (1) 25% energy consumption reduction, (2) 25.7% carbon emission reduction, and (3) 1% renewable energy generation.

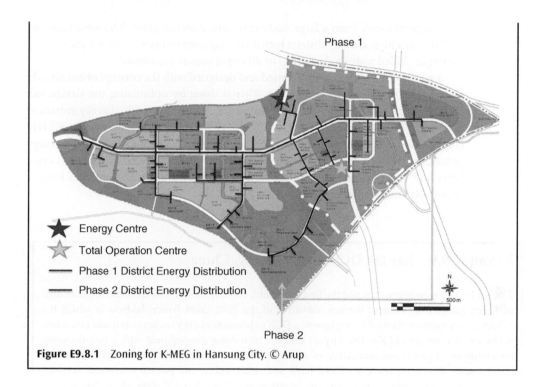

Phase 1

★ Energy Centre

☆ Total Operation Centre

— Phase 1 District Energy Distribution

— Phase 2 District Energy Distribution

Phase 2

Figure E9.8.1   Zoning for K-MEG in Hansung City. © Arup

9.5.2   District energy

Further improving overall system efficiency can be achieved by centralising energy conservation systems. With improved large-scale technologies becoming available, higher system efficiency at the community level by increasing economies of scale through application of District Energy systems can be achieved. In addition to higher efficiency, systems also benefit from centralising maintenance; hence, maintenance can be managed more effectively through a single professional entity. Moreover, prime space within individual buildings can be freed up for higher value usage since each building does not have to accommodate the necessary equipment within their own premises. The United Nations Environment Programme (UNEP) also promotes the use of such types of District Energy systems.[138]

In Hong Kong, the electricity use of air-conditioning accounts for 29% of total electricity consumption. A comfortable indoor environmental condition for the occupants of offices and other public spaces is a fundamental element in urban living. The energy efficiency of the air-conditioning system is therefore essential in achieving the energy saving targets of society as a whole. As a result, energy codes on controlling the design of air-conditioning and other building services systems are now in place in Hong Kong.

Amongst the different energy efficient systems, the District Cooling System (DCS) is the most sustainable solution for the planning of a new district, and is particularly suitable for developments with a high density or with clusters of buildings. Chilled water can be distributed to multiple buildings through an underground water pipe network from a large-scale centralised chiller plant. The advantage of the DCS in a high-density district is that the required infrastructure for the distribution of chilled water to buildings of different uses is minimised.

A successful DCS has to be planned and designed with the concept of enhanced system performance and economics. This is done by optimising the design of chilled water production and the associated distribution network, thereby enhancing energy efficiency and financial viability. The Kai Tak Development (KTD) DCS is one of the more recent examples (see Example 9.9), it has an energy-saving potential of 20% to 35% compared to traditional air-cooled air-conditioning systems and water-cooled air-conditioning systems, and users also enjoy a better quality and more reliable service.

Example 9.9   Kai Tak DCS in Hong Kong, China

The KTD DCS is an innovative, and the first of its kind, cooling method that has been implemented in Hong Kong. It was one of the key initiatives of the 2008–2009 Policy Address in which the HKSAR Government planned to implement a DCS to promote energy conservation and efficiency. KTD, situated on the old Kai Tak Airport site, is a mixed-use development with a non-domestic air-conditioned gross floor area (GFA) of more than 1.7 million m^2. The development consists of government offices, commercial offices, hotel and retail, public and private residential developments, community facilities, and transport infrastructure. The DCS will serve all the buildings in the KTD, except the domestic developments. In 2000, with the commencement of the feasibility study, system design, and then implementation this innovative and energy efficient system was finally realised. By using seawater for heat rejection, there is further energy saving, and with the removal of cooling towers more open space can be released to the public.

The KTD DCS comprises two separate plants: an associated chilled water distribution network, and customer substations. The south DCS plant room (South Plant) is situated under the old runway and serves the South Apron and Runway Boulevard of the KTD. The plant adopts a variable primary flow chilled water system and utilises direct seawater cooled heat rejection.

The north DCS plant room (North Plant), serving the North Apron, is situated at the northern end of Kai Shing Street and is adjacent to the Kwun Tong Bypass. Similar to the South Plant, the chiller system is a variable primary flow system with direct seawater cooled heat rejection. The main distribution chilled water pipe is a three-pipe ring circuit and is buried underground.

The DCS project was financed by The HKSAR Government and is being implemented in three phases (Figure E9.9.1). Phase I was mainly pipe laying and Phase II was construction of the chiller plant rooms, seawater pump house, and other associated facilities. Phase I and II works started in February and March 2011, respectively. Works in these two phases were scheduled to tie in with the earliest development in KTD, which includes the Kai Tak Cruise Terminal and public housing estates. Phase III, which began in July 2013, was the installation of additional electrical and mechanical equipment and the extension of pipes.

The South Plant commenced operation in February 2013, serving Kai Tak Cruise Terminal. Two months later, the North Plant (Figure E9.9.2) was in operation serving the new developments in the North Apron area.

The KTD DCS will reduce electricity consumption by approximately 35% compared to conventional air-cooled systems, on the basis of the design capacity of 284 MW. The estimated maximum annual saving of electricity is to be about 85 million kWh, equivalent to a cut of 59,500 tonnes of carbon dioxide emissions from the development.

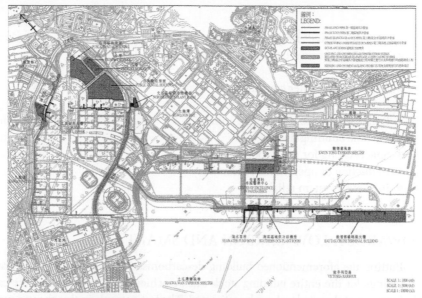

Figure E9.9.1 Phased Implementation of the KTD DCS. © Arup

Figure E9.9.2 North Plant and Chiller Installation. © Arup

Besides better energy efficiency, other environmental and planning benefits gained by adopting a DCS include the following:

- Reduction of water consumption
- Reduction of carbon footprint
- Reduced noise and vibration impact
- Flexible building design
- Space-saving for plant rooms
- Reduction of operational redundancy
- Improved system reliability

9.6   TOWARDS A LOW-CARBON AND SMART CITY

Putting the aforementioned building de-carbonisation strategies together, carbon savings for the entire building stock can be achieved. Taken together, the carbon footprint at the city level can be estimated. The last example in this chapter summarises recent building projects in which the authors are involved and suggests the level of carbon savings from the individual building that can be achieved (Example 9.10).

De-carbonisation of a city or making a sustainable city is more than just green buildings; energy, transportation, safety and security, the natural environment, people, and culture are also very important aspects that need to be included. Cities are required to be smart and resilient to external risks and must be responsive to the people's needs beyond just using eco-focused technologies.

Firstly, many design best practices and industry guidelines to help design buildings for carbon neutral performance have been mentioned in Chapter 8. Being eco-focused

---

Example 9.10   Carbon saving from buildings in Hong Kong, China

In many cities, building energy requires regulatory approval for building design and construction. Building codes and regulations spell out the minimum standard for building design. In Hong Kong, building energy codes (BEC) are regulated by the Electrical and Mechanical Services Department (EMSD). In recent years (2007, 2012, 2015), the BEC has been updated more frequently to follow industry trends more closely. BEC in Hong Kong covers four major categories: (1) air conditioning, (2) electrical, (3) lighting, and (4) lift/elevator installations. These categories comprise 80% of building energy use. Given the requirements, the maximum building energy use for a new building design can be determined and that design can be treated as baseline. The majority of energy use is from electricity and the carbon emission value from electricity production is recent years is around 0.7 kg CO_2/kWh in Hong Kong. The information, when combined, is helpful for determining the carbon footprint of the building sector.

To get an impression on how much recent building technologies have improved de-carbonisation, the following table (Table E9.10.1) summarises the carbon saving for several building projects. For common types of building, such as institutional, infrastructure, commercial, office, and private residential, the carbon saving for the proposed building designs range from 16.1% to 33.8% from the baseline designs of the reference year. All of these new developments demonstrate that carbon reduction in buildings is a big opportunity going forward.

Table E9.10.1   Carbon Saving for Recent Building Projects in Hong Kong

Building (District)	Total GFA (m²)	CO_2 Proposed Design (kg CO_2/yr)	CO_2 Reduction from Baseline	Baseline Ref Year
Institution (West Kowloon)	88,000	14,531,720	25.1%	2007
Institution with DCS (Kowloon Bay)	65,470	5,433,420	36.7%	2007
Institution (Pok Fu Lam)	5,160	591,057	19.2%	2012
Infrastructure Facility (Shum Shui Po)	21,388	3,080,490	21.8%	2012
Commercial (Causeway Bay)	43,259	7,488,418	27.6%	2012
Commercial (Quarry Bay)	94,427	10,198,893	32.1%	2012
Commercial (Central)	9,584	2,453,847	29.6%	2012
Office (North Point)	30,635	3,946,650	30.6%	2012
Office (Kowloon Bay)	82,096	12,502,378	33.8%	2012
Residential (Yau Ma Tei)	79,055	7,587,900	19.0%	2007
Residential (Yau Ma Tei)	105,067	10,738,100	17.3%	2007
Residential (Tsuen Wan)	99,757	11,395,600	19.9%	2012
Residential (Yuen Long)	59,366	5,614,200	16.1%	2015

means using smart technologies, such as Internet of Things (IoT) and Big Data, to harvest, analyse, and manage energy, water, and waste more sustainably.

Secondly, with the pressures of extreme climates and urbanisation, in the future cities will be required to be more resilient to unpredictable external events. Data analytics with large amounts of data allows us to monitor, correlate, predict, and provide alternative solutions to trends that might interrupt the normal functioning of a city, for example transportation, safety and security, the local economy, and housing.

Finally, residents in cities look to the local government to better meet their needs. To attract more talent by creating a better standard of living, a city should be responsive to its citizens in the areas of health, social welfare, community services, and education.

9.7   SUMMARY

To de-couple our daily living habits with carbon emissions is the ultimate goal of building sustainability. The de-carbonisation of buildings or even the entire community is technically feasible. The high-rise and compact cityscapes of East Asian cities have complicated the process. Yet, the advancement of zero carbon technologies coupled with smart solutions is able to make our buildings and energy infrastructure very efficiency. It is envisaged that with more reliable renewable technologies such as bio-fuel from algae available on the market, we could truly see eco-cities prevailing in East Asia.

Section 4
On people

Chapter 10
Space for people

10.1 INTRODUCTION

Cities are more likely to want better sustainability performance as a way of building up the local economy and community according to their own unique and special sense of place. The location-based urban design concept is prevailing.[139,140] This is generally a "bottom-up" process in effective city design. If the location is given more attention it is likely that the ecological footprint will be reduced, and the ecological features will be protected. Most cities treat economic development as the first priority, so making the place unique and improving the wealth of the community is the best way to achieve this. A city's heritage and its unique culture will form the foundation of a good economy, especially when a community has a strong connection with the local environment.

Cities provide opportunities for people to work as well as provide the necessities for living. The recent urban development of China has demonstrated the power of urbanisation to lift millions of people out of poverty. Many years of urban development has transformed the physical form of "modern" Asian cities beyond recognition, high-rise buildings have replaced low-rise houses and mixed-used complexes have replaced street markets. Yet, the space provided for people is far from helpful in providing a good quality of life. Among the most pressing problems faced is the ventilation issue, at the macro level in high-rise cities and at the micro level within residential units. The poor environmental quality of open spaces is seriously affecting the health and well-being of the population and has created a heat island around people's living spaces. Designers in Asia have been studying the issues and have applied innovative measure to alleviate the problems that cities have created.[141,142,143,144,145,146]

Building Sustainability in East Asia: Policy, Design, and People, First Edition. Vincent S Cheng and Jimmy C Tong.
© 2017 John Wiley & Sons Ltd. Published 2017 by John Wiley & Sons Ltd.

10.2   URBAN CONTEXT OF ASIA CITY

Major cities in Asia are megacities are characterised by compact and high-rise urban design with a densely populated living environment. The urban environment has many problems and some may pose a threat to the well-being of residents. Governments have endeavoured to improve the liveability of urban spaces but the journey towards sustainable development is still a long one for many city dwellers.

10.2.1   Liveability

Urbanisation has driven most people living in cities. The living environment, and its quality, varies significantly across different countries. "Liveability" is the sum of the factors that add up to a community's quality of life. This includes the built and natural environments, economic prosperity, social stability and equity, and educational opportunities, as well as cultural, entertainment, and recreation possibilities. In 2015, Vienna ranked number one as the most liveable city in the world. Through the process of urbanisation over the past few decades, Asian cities also created many quality living spaces that can be showcased worldwide. For example, Singapore is famous for its greenery,[147] and Hong Kong was once ranked as the most liveable city in the world by the Economist Intelligence Unit because of the vibrancy of its city life.[148] Yet, the liveability of a city can only be truly experienced by its citizens. The design of the urban environment has evolved over the years to be able to offer a quality lifestyle for people to enjoy and it forms the key social asset for a community.

The quality of space in city can be defined by many social, cultural, and environmental factors, and it is greatly affected by the perceptions of the people to the urban context and the environments formed. The buildings, the open spaces, and the greenery can all determine the quality of the environment, which can be measured qualitatively by the response of the people and by the local environmental conditions created by the buildings or the microclimate. How to create a favourable microclimate that attracts people to use the space is the key to successful urban design.

In Asia, governments and design professionals have tested and verified many different design concepts in order to improve the built environment associated with high-density living. The challenges they face are similar, yet the solutions vary. This chapter will look at the problems and identify viable solutions that will help to improve the quality of life in urban living.

10.2.2   A compact and vertical city

Most Asia mega-cities favour a compact city design as it helps to provide the urban sustainability achieved by dense, mixed neighbourhoods, which in turn make the city more liveable. A report prepared by the Centre for Liveable Cities (CLC) in Singapore has indicated that Singapore and Hong Kong, despite having the highest density of buildings in the world, can also provide a high quality of liveability at

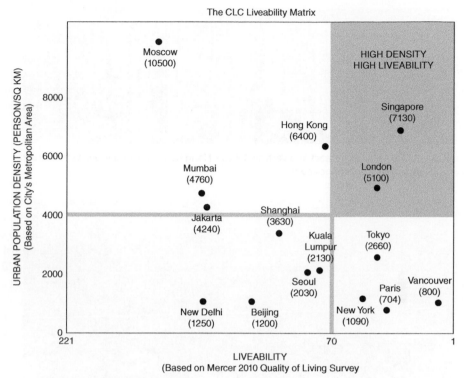

Figure 10.1   CLC Liveability Matrix. © Arup

the same time.[149] Both cities have embraced a high-rise urban fabric model and high-rise public housing development has been used by Hong Kong and Singapore to meet their housing requirements and to make their high-density environments more liveable.

Hong Kong is one of the most densely populated cities in the world, with a population of 7 million on a mere 1,100 km^2 of land in total, and it has developed into unique urban high-rise and high-density environment (Figure 10.1).

In the home, there is less living space per person among all of the major cities in Asia (as little as 10 m^2 per person; compare 20 m^2, 25 m^2, and 30 m^2 for Tokyo, Shanghai, and Singapore, respectively). Gradually, taller and taller buildings get built and a vertical cityscape is formed. One of the challenges is how to maintain a liveable but very compact built environment. The solution is by providing quality open space in urban design.

Urban open space is crucial for city sustainability. The high density of buildings does pose a challenge but better planning, design, and management can reduce their impact by making the living and working conditions less crowded. The perception of the crowdedness depends on the surrounding environment.[150,151]

Similarly, the development of high-rise buildings in tight spaced sites poses challenges to architectural planning, design, and construction in terms of achieving optimal indoor comfort and healthy living conditions. The creation of quality outdoor and indoor space becomes imperative for people living in cities (Figure 10.2).

Figure 10.2   A Tale of Two "Compact and Vertical" Cities – Hong Kong and Singapore. Left: © Arup; Right: © TrongNguyen/Shutterstock.com

Figure 10.3   Urban Environment of Hong Kong. © Arup

10.2.3   An undesirable building environment

In Hong Kong, the planning and design of buildings within a development boundary is controlled by building regulations. However, the requirements were first formulated in the 1950s and 1960s when the urban context was very different. With building density increasing in build-up districts over the next few decades, building design and the built environment based off such outdated requirements will not be effective enough and in some instances will not be able to safeguard health or safety. The outbreak of Severe Acute Respiratory Syndrome (SARS) in 2003, a viral respiratory disease that is spread by close person-to-person contact, was a wakeup call to the deteriorating conditions of Hong Kong's compact built environment (Figure 10.3).

Amongst the many reasons that contributed to the outbreak was the easy transference of bacteria between the different floors of the buildings in the closely packed buildings and the provisions of re-entry in building. Many studies were conducted to determine the contributing factors. Environmentally, it was a result of inferior outdoor ventilation and lighting conditions. It also created negative psychological effect.

10.3   THE QUEST FOR A QUALITY BUILT ENVIRONMENT

In response to the request from society for a better living environment, many governments in Asia have taken action to ameliorate the design of new developments or major re-developments. In Hong Kong, the government conducted a study to identify the areas requiring improvement.[152] The study looked into the limitations of current regulations and recommended some guidelines on planning and building design that could enhance the sustainability performance of buildings and its surroundings. In the urban context of Hong Kong, the key control areas are building separation, building setback at street level, and more greenery.

Other countries have also developed similar code of urban planning and building design. In Singapore, the Urban Redevelopment Authority (URA) plays a major role in developing a quality built environment. The URA also set up the Centre for Liveable Cities which was established to be a place of continuous learning and for the development of knowledge regarding liveable and sustainable cities. A result of the development control, urban design plans, and guidelines, was that the urban fabric now features pockets of green and blue spaces, park connectors, public spaces, sheltered walkways, better access to services and places of work, improved building form and separation, sky-rise greenery, building balconies, improved night-lighting, and outdoor dining areas. Within existing and upgraded public housing estates, the Housing Development Board in Singapore ensures that public playground spaces for three family generations, community facilities such as precinct pavilions, child care and elderly centres, green spaces, well connected communities, and housing with sustainable features are made available through programmes such as Remaking Our Heartland. At the building level, the Building Control Authority uses a combination of mandatory building control regulations, voluntary Green Mark certification (for a higher rating), and financial incentives to drive the adoption of sustainable features in both new and existing buildings. This has already driven the introduction of lush greenery, natural ventilation, and low carbon emitting materials that directly benefit the living environment.

In China, a planning guideline for sustainable urban design was implement in 2011. The challenges of high-density and high-rise building design are common and similar approaches were adopted in Asia.

10.3.1   A novel planning framework for the environment

A planning system was developed in Hong Kong, which integrates the urban map and local wind conditions with building design (Figure 10.4).

Over the past few years, the HKSAR Government prepared an urban climatic planning map for the entire territory. The system separates the planning into two levels – the territory wide level and the project level. At the territory-wide level, the outline zoning plan (OZP) of a district is prepared. Together with the planning standards, it controls the planning parameters for the district, such as building height, setback requirements, and building bulk. At the project level, the developer is required to carry out an Air Ventilation Assessments (AVA) as well as a micro-climate study to optimise the site planning and urban design such as greenery, and incompliance with the OZP.

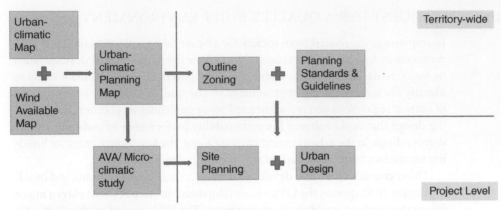

Figure 10.4   Planning Process for a Good Wind Environment. © Arup

Figure 10.5   The Focus of Urban Climatic Map Study: Reducing Ground Coverage and Increasing Greenery. © Arup

10.3.2   The urban climatic map

To resolve the long-term planning requirements for a better built environment in Hong Kong, a study to develop a city-wide urban-climate map was conducted.[153] The map shows the various climateopes, which is the classification system for the climatic characteristics of ventilation and urban heat islands (UHIs). Hong Kong's Urban Climatic (UC) Map provides information for strategic and district planning when outlining zoning plans and uses a scale of 1:2000. Because Hong Kong is a high-density city, the study's instructions and recommendations focused on reducing ground coverage to improve air ventilation and pedestrian comfort, and on improving greenery coverage in urban areas (Figure 10.5).

Based on the above reviews, it was discovered that UC Map helped to elaborate further on planning instructions and recommendations. As the goals of climate-sensitive planning need to be pursued in order to improve living conditions and its quality relative to bio climate, UHI, urban air ventilation, and the air quality situation, the main planning recommendations must focus on the following aspects: reducing the urban thermal load; controlling building volume and sealing areas; improving urban dynamic potentials; preserving, maintaining, and improving the existing urban ventilation paths and networks of the city, and charting new air paths if necessary; preserving, maintaining, improving, and respecting the cold air production and drainage areas of the countryside and vegetated hillsides near the urban areas; preserving, maintaining, improving, and respecting the land-sea breezes; preserving, maintaining, and improving urban greenery; and reducing the release of air pollutants, greenhouse gases, and anthropogenic waste heat. Thus, the planning strategies include four aspects: albedo, vegetation, shading, and ventilation; and they are aimed at improving the urban climatic environment offer important lessons for sustainable urban development in the future.

It has only been a short while since the launch of the UC Map and efforts for sustainability are still required. Planners should make sure they understand the UC Map and any planning instructions from urban climatologists. Balance and optimisation must also be sought for all concerns and issues and a professional team should be formed to monitor the effectiveness of the application. The UC Map team will continue to collect and evaluate further data, and refine their scientific base as knowledge improves. They will also update and improve the map to cope with the changes in urban morphology (Figure 10.6).

Figure 10.6   Urban Climatic Map of Hong Kong. © Arup

10.3.3   Air ventilation

The outbreak of SARS in Hong Kong triggered an investigation by the Planning Department into the air ventilation performance of Hong Kong's urban environment. It was part of the actions recommended by the Clean Team which was formed after the SARS incident. The study developed a new tool called the Air Ventilation Assessment (AVA) for the assessment of the performance of ventilation in the urban environment. It is more scientific and objective to "measure" the deficiency of urban design, such as the "wall effect" or the "urban canyon phenomenon" by means of computer modelling or wind-tunnel measurement. The AVA is now part of all technical assessments for any planning submission under Planning Ordinance Section 16.

The implementation of the AVA in Hong Kong has provided the industry with useful and powerful information on the status quo and the scenarios of effective planning in the future. The data available at the moment is quite diverse among the different areas of Hong Kong due to the variation in building design and density. Velocity Ratio (VR), which measures the local velocity compared to the reference velocity, is the performance indicator for comparison. Typical VR in different districts in Hong Kong is summarised in Table 10.1 for reference.

In the experimental wind tunnel study from the Planning Department of the Hong Kong Government, the mean VR measured in the scaled model of the following areas is tabulated below. The low-density urban context was found to have the highest mean VR which then decreases with building density. Example 10.1 describes the CFD technical development to study the urban AVA.

10.3.4   Microclimate and landscape design integration

For the high-rise high-density urban environment of most Asia metropolises, creating a positive microclimate for the built environment is important. Many studies in the region have confirmed people's favourite conditions for outdoor environments. Strategies concerning the designs of streets, buildings, and open spaces have evolved. In particular, the hot and humid climate of South East Asia has made thermal comfort the key factor in design. Design strategies on enhancing

Table 10.1   Typical Velocity Ratio in Different Districts in Hong Kong

Urban Context	District	Mean VR
Low density	Wong Chuk Hang	0.22
	Tuen Mun	0.19
	Sha Tin	0.20
Medium density	Tseung Kwan O	0.18
	San Po Kong	0.15
	Tsuen Wan	0.15
High density	Mongkok	0.16
	Causeway Bay	0.16
	Tsim Sha Tsui	0.14

Example 10.1   Technical Development of CFD Approach for AVA in Hong Kong, China

In order to capture the more realistic flow patterns within an urban context, an advanced Computational Fluid Dynamics (CFD) technique is one of the more reliable and common tools used to study air ventilation in the urban environment. Various international wind CFD guidelines across different regions, such as the COST Action C14 for European countries, are usually referenced. However, the urban morphology for these countries is relatively low-rise and of such low density that the guidelines may not be fully applicable in cities such as Hong Kong which has a high-density context.

The Chinese University of Hong Kong, the Hong Kong University of Science & Technology, Tohoku University, and Tokyo Polytechnic University jointly carried out a research study with the intention of investigating the applicability of an advanced turbulence model, or Detached Eddy Simulation (DES), to model a dense urban environment. A comparative study between the measured data in wind tunnel tests and the CFD results with DES and standard k-ε model was conducted (Figure E10.1.1).

The DES was found suitable for studying the pedestrian wind environment in the dense building context (Figure E10.1.2). It could predict more appropriately the velocity ratios at individual test points than those provided by the standard k-ε turbulence model.

Figure E10.1.1   CFD Study of Air Ventilation in Urban Development. © Arup

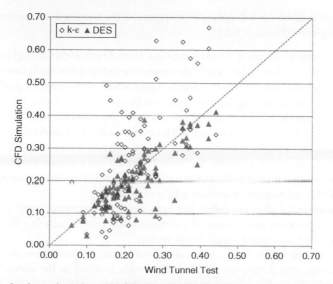

Figure E10.1.2   Study on the Urban Wind Environment. © Arup

thermal comfort by providing shade, greenery, and enhancing air movement are becoming more prevalent in design.

The term microclimate refers to the local modification of the general climate by the immediate surroundings of a smaller area including topography, ground surface, plant cover, and any man-made forms in the vicinity. Microclimate studies integrate the latest proven scientific technologies, including computational fluid dynamics simulations, wind tunnel modelling, daylight simulation tools, and dynamic thermal modelling to study microclimate.

Built Environment Modelling (BEM) is an essential tool for multidisciplinary environmental design (i.e., sky-views, solar design, ventilation, energy-use). In recent years, a sharp increase in the demand for high performance building design has been noticed. This demands an integrated approach, even seemingly simple questions such as "if we provide shade from the sun, will we block out the wind?", or "if we change the aspect ratio of a building, will it reduce energy consumption?" requires the designer to handle various environmental parameters. The idea is to provide a flexible interface to allow designers to integrate results from a range of specialised environmental modelling software tools and present the results graphically to enhance an iterative design process between the various streams of designers (architects, engineers, planners, etc.).

Examples are introduced in which a closed-loop feedback system between parametric architectural design and BEM is established: output from BEM (lighting, thermal comfort, and energy use) is used as direct input to a parametric code that generates the architectural form; the architectural form is used in turn as the input to the BEM platform. The form evolves in this iterative design process and the final convergent product represents an optimised form that reflects the needs of the local microclimate (Example 10.2).

Example 10.2   Microclimate for housing design in Hong Kong, China

Microclimate studies effectively follow the different stages of the housing development project from inception to completion (Figure E10.2.1).

1 Planning and conceptual design - at this planning stage, microclimate studies are deployed to compare different site planning options and to identify the appropriate strategies that enable favourable access to wind, daylight, and prevention of excessive solar heat gain, as well as provision of a good view.
2 Scheme design - during this stage, microclimate studies assist to determine the suitable orientation for the residential blocks to capture more wind in the common areas such as corridors, lifts, and entrance lobbies as well as the optimal locations for different outdoor spatial allocations such as sky gardens, empty bays, outdoor activity areas, and refuse collection points.
3 Detailed design - microclimate studies are applied to study and optimise the design of interior spaces and the incorporation of building features (e.g., wing walls) to enhance natural ventilation, optimise daylight, and minimise solar heat gain, and so on.
4 On-site measurement - after the design has been set down by microclimate studies, on-site measurements are carried out to cover all the major design parameters that the microclimate studies provided for the environmental design aspects.
5 Post-occupancy stage - a site survey is be carried out to collect information from residents in terms of environmental design; the result would provide feedback to the microclimate studies model concerning its studied parameters and local criteria.

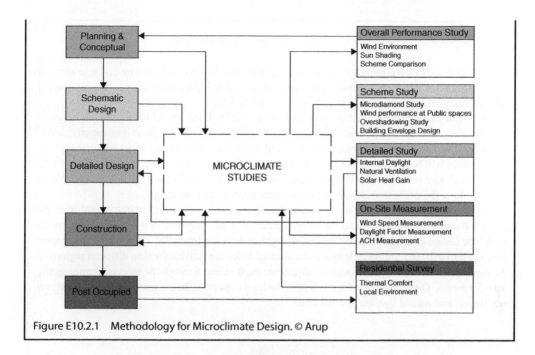

Figure E10.2.1   Methodology for Microclimate Design. © Arup

Firstly, BEM is an extension of the now popular Building Information Modelling (BIM) system – it uses the three-dimensional data from BIM to infer the environmental performance.

Secondly, it is an integrated technique – individual parameters (such as wind modelling or solar modelling) have been set in a piecemeal way for years. The recent availability in computational power and three-dimensional graphical techniques has allowed users to combine the information and produce enhanced design-oriented data.

It is time of functional, rational, and information driven design. The second point is of particular importance. The design of a complex system, for example, a building, consists of a series of design decisions, where the benefits of individual strategies are traded off against one another. The vast range of environmental parameters becomes truly useful when they can be integrated and considered in relation to each other (Example 10.3).

10.4   REDUCING THE URBAN HEAT ISLAND

Over the years, we have witnessed high-density urban development all over the world. By 2050, there will be more than 20 new mega-cities in Asia. If these new cities are not designed sustainably, it will lead to significant environment deterioration and induce hotter temperatures (Figure 10.7).

An Urban Heat Island (UHI) effect is an urban area that is significantly warmer than the rural areas due to urban development. This effect can be commonly observed in the cities all across the world. The intensity of UHI can be used to describe the cumulative impact of a city, and the higher the value of UHI the higher the impact.

Example 10.3   The South Beach project in Singapore

The South Beach Development in Singapore includes two towers (a 45-storey office tower and 34-storey hotel and residential tower) and a semi-outdoor retail space covered by an environmental filter (canopy) at around 300 m in length (see Figure E10.3.1).

The South Beach development is intended as an iconic landmark on which a complete range of sustainability strategies are showcased. The development is mixed-use, providing the city centre with high-quality office space, premium hotels and residences. Two major towers and accompanying podiums will be added to the site, and will be architecturally integrated with existing heritage buildings. A key feature of the design is the "environmental filter" canopy that covers the open space, linking the two towers, covering the podiums heritage and buildings.

Sustainable design is not simply about reducing energy or eye-catching environmental features, other features such as water conservation and sustainable building materials must also be considered. The design addresses issues of sun, wind, light, and rain individually. These elements are then combined to deliver a comfortable space that is suitable for the activities within different regions of the canopy. Furthermore, such design requirements must be met through the most environmentally friendly means. Only measures which will have the least impact on the environment, such as natural ventilation and natural lighting, are selected.

Figure E10.3.1   South Beach Development, Singapore. © Arup

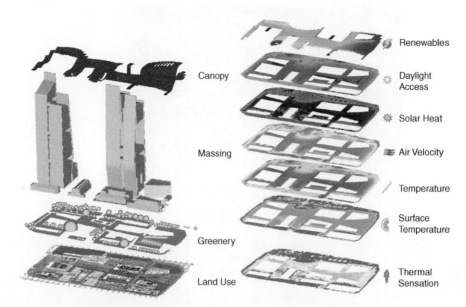

Figure E10.3.2   Building Environmental Modelling of the Project, Singapore. © Arup

   Central to the design strategy, the canopy must act as a modifier to the semi-outdoor environment within. The resulting environment should be tailored to the "destination location" nature of the site. Appropriate levels of wind, solar radiation, air temperature, and rainfall must be provided for the respective areas within the canopy, such as roof gardens, pedestrian walkways, and plaza areas.

   A holistic approach with assistance from BEM is a key to the sustainable design of the towers and secondary buildings. The design integrates microclimate within this canopy with all environmental aspects to inform an optimised solution that provides daylight, thermal comfort, rain protection, and renewable energy. All aspects of the environmental design are incorporated in BEM and the BEM builds a performance map layer (Figure E10.3.2) for each of the parameters: space use, massing, solar radiation, ventilation performance, wind availability, air and surface temperature, greenery, daylight, rain protection, and thermal comfort index. This integrated approach provides the design inputs to optimise the shape, porosity, materials, and details of the canopy.

   As a result, the canopy design represents an innovative solution to the problem of semi-outdoor environmental comfort in Singapore; it provides an appropriate level of protection from rain and sun while introducing desirable diffuse light and cool breezes to the users within:

- Air Ventilation - to remove heat and bring in fresh air, the canopy form is aerodynamically optimised to induce and accelerate air into the open space via the canopy openings and the openings are included to maintain stack ventilation during windless days;
- Sun Shade and Daylight - while maintaining good daylight access, the canopy panels provide shade to give sun protection at the pedestrian level for human thermal comfort;
- Rain Protection - with rain every three to four days, rain protection in the sub canopy zone can increase the usable area for the space;
- Renewable Energy - the canopy is optimised for harvesting solar energy, where optimised panel arrangements for cost effectiveness are achieved.

Figure 10.7   Effect of Greenery on Reducing UHI. © Arup

Among the world's top mega-cities, Hong Kong is one of the most densely populated places in the world. In 2014, the population density can be as high as 57,250 persons per km² in some districts.[151,154]

Owing to the intensive urban environment, Hong Kong is suffering from a UHI effect. The urban environment has been found to be significantly warmer than in the rural areas. The Hong Kong Polytechnic University used infrared satellite imaging and found that the temperature difference between urban and rural areas was around 5 °C on summer nights and 7 °C on winter nights, while the maximum difference was as high as 12 °C. The UHI effect leads to an uncomfortable living environment, heat stress, can generate health problems, increases energy consumption, and can cause knock on economic problems. In short, UHI creates a poor living environment.

To confront such a challenge, it is vital to improve the conventional urban planning process. The new approach needs to integrate the climate and environmental considerations into the city planning to provide an opportunity to design in a responsive way (Table 10.2).

UHI modelling could be one of these approaches to improve the conventional urban planning process. With the support of the latest technology and advanced

Table 10.2   Factors in UHI design

Design Factors	Design Considerations
Solar Storage	Material selection & allocation, local shading
Wind Speed	Building separation, building permeability
Tree Coverage	Tree allocation, walkway strategy
Green Coverage	Greenery ratio
Water Feature Coverage	Landscape design
Sky View Factor	Building geometry

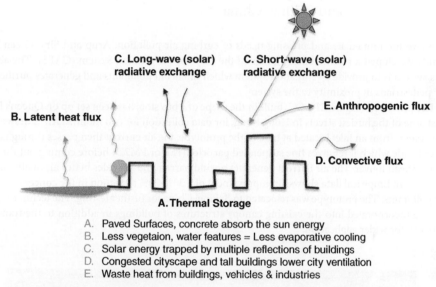

C. Long-wave (solar) radiative exchange

C. Short-wave (solar) radiative exchange

E. Anthropogenic flux

B. Latent heat flux

D. Convective flux

A. Thermal Storage

A.   Paved Surfaces, concrete absorb the sun energy
B.   Less vegetaion, water features = Less evaporative cooling
C.   Solar energy trapped by multiple reflections of buildings
D.   Congested cityscape and tall buildings lower city ventilation
E.   Waste heat from buildings, vehicles & industries

Figure 10.8   UHI Design in Hong Kong. © Arup

computational techniques, a new modelling approach can be developed to predict and visualise the heat distribution and wind flow of the city. The information can then be utilised to facilitate better city planning including identifying potential problematic areas and optimise the design of the city to facilitate more wind penetration and greenery. The UHI model includes modelling of various environment factors, and the results are in terms of temperature rise compared to the supposed air temperature of the virtual rural areas in that location. The key environmental factors include wind speed, solar heat distribution, greenery distribution, tree coverage, water feature coverage, and the sky view factor (Figure 10.8).

10.5   STREET CANYON EFFECT – ROADSIDE AIR POLLUTION

Busy road traffic in a densely populated city causes poor roadside air quality and it is a threat to public health. Due to the density of high-rise buildings, a common feature is that many narrow street canyons are created. The change of flow regime, known as the Street Canyon Effect, results in the accumulation of pollution and is one of the biggest problems in modern cities.[155] Due to long exposure, pedestrians can suffer from serious health problems. Most of these types of cities face the challenge of achieving an acceptable air quality level on the street. In Beijing during 2013, it was reported that there were 58 "very unhealthy" days which had an Air Quality Index (AQI) reading of higher than 200.[156,157] An innovative solution was created to combat local roadside pollution and it is explained in Example 10.4.

Example 10.4   City air purification system in Hong Kong and Beijing, China

To serve the immediate and pressing needs of curbing air pollution, Arup and Sino Green have jointly developed a patented design, named the City Air Purification System (CAPS). The aim of this system is to provide a ventilation system which filters out pollutants and generates purified air for pedestrians in proximity to the system.

The prototype (Figure E10.4.1), built in the shape of a bus stop, has been set up on Queen's Road East, one of the busiest streets in Hong Kong, for data collection for two months. Air is drawn into the system from an inlet located at base of the prototype, the air current then passes through a bag filter inside which it removes fine suspended particles (FSP or PM2.5) before coming out through an overhead louver. The air current generates a wind barrier and provides better air quality inside the system. Empirical data shows an improvement of a 30%-70% reduction in the concentration of air pollutants. The prototype was relocated to Beijing, China for further testing. The technology can also be incorporated into the existing canopy structures of buildings in addition to the transport network for wider application.

Figure E10.4.1   City Air Purification System Testing at a Busy Roadside at Wan Chai, Hong Kong. © Arup

10.6   RIGHT OF LIGHT

It is a fundamental right by law for everyone to enjoy access to natural light. Many building regulations include such a principle. In China, there are requirements in the building regulations which safeguard a minimum of one hour on winter days for habitable space to receive sunlight. Any new development or redevelopment nearby must not infringe on this right, if it does so the approval for development will not be granted (Figure 10.9).

In Hong Kong, however, this right has been abolished as the city has changed into a high-rise urban context. The requirements only provide very limited access to the sky. In fact, the survey has shown that some windows facing the "re-entrant" could hardly see the sky at all. This results in very unfavourable lighting conditions in the habitable areas, that is, the living rooms, bedrooms, and kitchen. The law, which is meant to safeguard the basic right to natural light, allows for unfavourable conditions due to poor building design.

10.7   HEALTH AND WELL-BEING

Apart from the minimal environmental impact in terms of resource consumption and the natural environment, sustainable buildings should also be able to provide a healthy and environmentally friendly environment for the occupants. To create a healthy indoor built environment, the building should have the provisions to facilitate holistic human's wellness aspects: air quality, water quality, nourishment, light, fitness, comfort, and mind (Figure 10.10).

Starting with architecturally passive design, buildings should get the most use out of nature in terms of natural light, ventilation, solar warming, and so on.

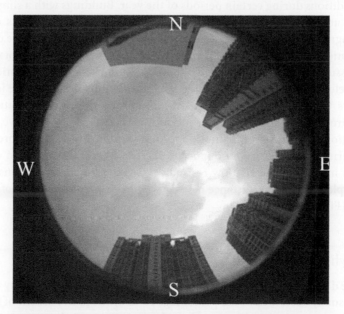

Figure 10.9   Sky Component of Typical Development in Hong Kong. © Arup

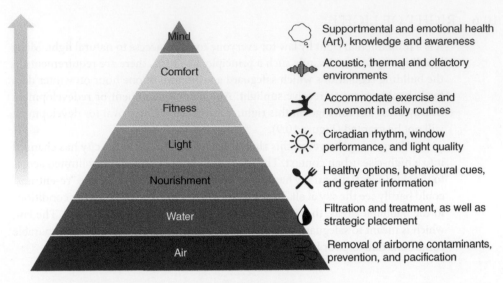

Figure 10.10   Well-Being Components of Healthy Building Design. © Arup

10.7.1   Natural ventilation

Sufficient ventilation not only can maintain health, but can also enhance the level of thermal comfort. The air supply system should be able to provide large amounts of fresh air for the occupants, and a free cooling method should be provided to induce higher amounts of fresh air when the outdoor conditions are moderate.

Natural ventilation is encouraged for regions with moderate climatic conditions during certain periods of the year. Buildings with a simple floor lay-out are good for applying natural ventilation. Building physics analysis and design are required to check the prevailing wind conditions at the development location, as well as in conducting an air flow simulation to obtain the wind pressure along the entire envelope. With wind pressure distribution along the envelope, building physics engineers can recommend the suitable locations and sizes to have window openings, according to the heat generation rate and desired room conditions of the indoor units. To facilitate effective natural ven-tilation (i.e., cross-ventilation), opposite window openings are preferable. The internal layout is also an important factor in natural ventilation effectiveness and in minimising air flow resistance. Key design considerations include the simple air-pathway, distance between opposite openings, depth of unit, and so on (Figure 10.11).

10.7.2   Daylight for habitation

The daylight factor is a measure of the subjective daylight quality in a room. It is usually described as the percentage ratio of the total amount of illumination falling onto an area compared to the instantaneous horizontal illumination

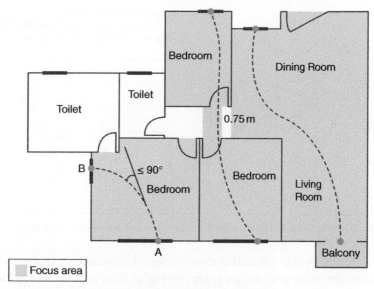

Figure 10.11   Cross-Ventilation Design in Residential Buildings. © Arup

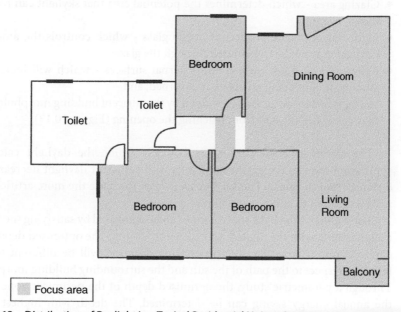

Figure 10.12   Distribution of Daylight in a Typical Residential Unit. © Arup

from a complete hemisphere of the sky, excluding direct sunlight (i.e., an overcast sky).

In general, the focus is put on rooms where the demand for access to natural daylight is essential. For example, the living/dining room and all bedrooms in domestic buildings; and the perimeter zone for commercial buildings (Figure 10.12).

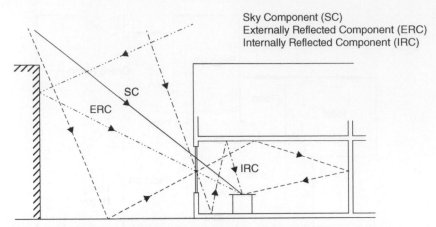

Figure 10.13   Daylight Factor Calculation Methodology for High-Rise Buildings. © Arup

With daylight calculation there are some key parameters that predetermine the daylight factor of an indoor space (the higher these parameters are, the higher the daylight factor that can be achieved):

- Glazing area - which determines the potential area that skylight can reach the indoor space;
- Light transmittance properties of the glass - which controls the amount of daylight that can be transmitted through the glaze;
- Reflectance of both internal and external surfaces - which will increase the amount of light being reflected or absorbed; and
- Sky view factor - which is determined by the adjacent building morphology that may shield the skylight from reaching the opening (Figure 10.13).

The aforementioned parameters basically define the daylight calculation approach; however, the extent of an area with sufficient daylight decreases with distance from the glaze. The further away from the glaze the more artificial light will be required to enhance the luminance level.

The depth of a daylight control zone can be optimised by satisfying the lighting design criteria with the highest annual energy saving. The optimised depth of the daylight control zone (i.e., the penetration distance) will be different for each façade in respect to the path of the sun and the surrounding building morphology. Through a parametric study, the optimised depth of the perimeter zone against the annual energy saving can be determined. The depth with highest annual energy saving can then be selected to be the optimised depth of the perimeter zone (Figure 10.14).

10.7.3   Water quality

Clean drinking water is a primary requirement to maintain human health; however, people commonly lack the knowledge to ensure healthy water quality. There are also more than the expected numbers of contaminants people need to

Energy Saved vs Daylighting Control Zone Depth:

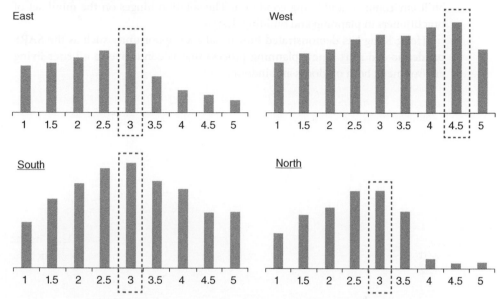

Figure 10.14　Energy Saving Extent against Daylight Control Zone Depth (m) for Typical Office in Hong Kong. © Arup

tackle in order to ensure that safe drinking water is delivered to the end-use water tap. Generally, there are four known categories of water contaminants that can be identified and thus filtered or have their contents reduced; namely, fundamental water quality, inorganic contaminants, organic contaminants, and agricultural contaminants.

Fundamental water quality is the basic indicator to ensure water quality, and it requires low turbidity and an absence of coliforms. For inorganic contaminants, various metals (e.g., lead, mercury, nickel, and copper) need to be kept at very low levels to avoid the risk of high blood pressure and kidney problems. Generally, reverse osmosis (RO) systems are used to remove these dissolved metals. Regarding organic contaminants, they are found in ground and surface water, with common contaminants being polychlorinated biphenyls (PCBs) and vinyl chloride. They may cause adverse health effects including cancer, immune deficiencies, and nervous system difficulties. Activated carbon filters are applied to remove these harmful substances. Pesticide compounds from agricultural activities may enter the water supply and lead to adverse health effects such as kidney problems and reproductive difficulties. If pesticides are detected, carbon filters are needed to remove these chemicals.

10.8　SUMMARY

In Asia, the development of the high-rise and high-density built environment is a matter of necessity rather than choice. The rapid process of urbanisation has created a huge pressure on housing and land. Yet, the experiences in Asia have demonstrated that high-rise compact cities are not necessarily equal to inferior built environment and living condition. The solution hinges on the mind-set of practitioners in planning and building design.

Hong Kong has demonstrated how to take an opportunity such as the SARS pandemic and start a new planning process that is conducive to a better living environment, both outdoors and indoors.

Chapter 11
Community making

11.1 INTRODUCTION

A community is any group of people living together in one area. Any town or village can be classified as a community, whereas cities often have smaller communities defined by particular areas. This paradigm of communal networks and shared social understanding has been applied in multiple cultures throughout history. Asian cities share a vibrancy of dynamic city life. Variations in culture have provided opportunities and challenges of making communities sustainable, both socially and environmentally. Development framework or model prevailing in western countries has been applied and texted. Their experiences are valuable for other cities worldwide that are undergoing a similar process of urban transformation.

Generally, a sustainable community includes the following objectives: well-being, low carbon emissions, convenience, social networking, and economy. Given the high density of urban living conditions, the design of sustainable housing developments has become of paramount importance to the quality of living and the well-being of people. The government is responsible for developing public housing, which affects a large area of land use and a many people. Thus, it is important for the government to lead developments with a comprehensive and sustainable strategy.

Aside from housing developments, a community uses many buildings types for various purposes. A community, town, or district is designed through thoughtful master planning, where multi-disciplinary measures and aspects are included. As with green building certification programs, the practice of planning and designing green/sustainable communities are established. This chapter describes how sustainable communities are formulated and practiced in Asian cities.

Building Sustainability in East Asia: Policy, Design, and People, First Edition. Vincent S Cheng and Jimmy C Tong.
© 2017 John Wiley & Sons Ltd. Published 2017 by John Wiley & Sons Ltd.

11.2   SUSTAINABLE COMMUNITY

Issues on the sustainability of a community are multi-faceted. When we strive to ensure the sustainability of our society, we should make every effort to improve the living environment for our people and respect and preserve the cultural heritage of society. Culture affects lifestyle, and therefore one's carbon footprint. The concept of a low carbon community has been proposed to accelerate a prosperous low carbon economy. Various low carbon technologies, strategies, and lifestyles, involving community planning, advanced green building technology, recycled water and waste management systems, change of low carbon living modes and energy-related behaviour are being developed in new Asian cities.[158] The goal is to encourage neighbourhood design and land use planning methods, which will reduce environmental impacts while keeping community liveable. Many new local planning codes/guidelines have been developed in various regions, which has helped to transform communities. Figure 11.1 and Table 11.1 show a collection of recent projects on this topic.

Even in professional circles on sustainable development, there is no universal definition of a sustainable community. As different locations face different opportunities and constraints for developments, a different set of measures and criteria of a sustainable community is used to characterise the design approach. Given the uniqueness of individual community development, the following

Figure 11.1   Culture Diversity in Asian cities. Upper Left: © echoblue/Shutterstock.com; Upper Right: © Arup; Lower Left: © Tze Ming/Shutterstock.com; Lower Right: © Arup

Table 11.1   Leading Sustainable Communities in East Asia

Community	Low Carbon Characteristics and Measures
WKCD, Hong Kong	Art and culture facilities; urban park
Changxindian, Beijing	Low carbon planning, new institutional arrangement
EKEO, Hong Kong	Another premier Central Business District, a removed airport area
SongDo City, Incheon	International Business District on the waterfront
Cheonggyecheon River, Seoul	Restoration of a river, formerly a highway

dimensions contain elements for developing a widely accepted sustainable community:

- Social/Culture/Governance - identity, customs, traditions, heritage, engagement, recreation, education, safety, respect, care, physical and psychological wellbeing and health, government organisation, law and order, representation, accountability, and security.
- Environment/Ecology - air, water, waste, energy, emission, pollution, habitat, build-form, transport, nature, ecosystems, biological diversity, and renewable resources.
- Economics/Living - housing, consumption, production, sourcing, investment, business ownership, employment, workforce, and training.

The aim is to design, create, and redevelop a sustainable community, and it is essential to take an integrated approach to balance human, natural, and financial resources regardless of the size or scale of a community. People involved in the process should take a long-term view that creates an opportunity for communities to be resilient to present and future needs. Citizens, leadership, institutions, and businesses should work together to express their priority among the elements listed. They must also collaborate to find workable solutions to make a sustainable community a reality for all parties.

11.3   COMMUNITY-BASED DESIGN

An urban design for an equitable community must focus on many social and cultural considerations beyond the requirements of environmental and economic benefits. Community driven urban designs employs creative tactics focused on social impact, civic dialogue, and grass roots place making. Also known as a "bottom-up" approach, community-based designs have gained momentum in the twenty-first century due to a variety of economic, governmental, social and technological factors. It empowers people and communities to participate, which is a process that puts people or end-recipients at the centre. This will allow the community to be integrated and engaged in every step of the process. In addition, this could also avoid the possibility of community members feeling alienated by a history of development projects.

11.3.1   Cultural aspect (social)

Culture is a crucial component for any community development. Re-vitalization of local traditions is key to many development projects in East Asian countries, which may face many hurdles from various stakeholders. Recently, the restoration of a culturally important river in Seoul, Korea has taught us a key lesson. A river can be recovered and restored even in a large and dense central business district, with multiple positive effects. These include providing a natural habitat for wild-life, preserving cultural heritage, public access to nature, flood control and micro-climate regulation. The river restoration also promotes more sustainable transport forms on roads, and cars removed allow more space for the river and its surrounding parks. The success of Seoul's Cheonggyecheon river restoration has had massive ripple effects: in East Asia and North America, cities are studying the project to gain its benefits for ecology, environmental quality, and urban sustainability.

11.3.2   Placemaking (environment)

People are the basis of all communities, and these establishments are often surrounded by nature. Therefore, it is important to consider the environmental impact during the development of a community. Placemaking refers to connecting people with places, where humans and nature coincide together.[159] Placemaking advocates maximising the quality of life and environment through the engagement of citizens. It is important to establish a love of nature and to recognise our impact on an ecosystem before an effective approach to incorporate strategies for developing a community can embrace an environment. Previous chapters have covered a lot of design strategies that can improve the impact of the development to the environment. Furthermore, it is more important for decision makers to prioritise various competing aspects to balance the values for both people and the environment. Placemaking involves multi-disciplinary corporation to bring forth multiple causes, such as resilience, engagement, wellbeing, preservation, and economic development for the people, city, and environment. The transformation of Kowloon East (comprising Kai Tak Development, Kwun Tong and Kowloon Bay Business Areas) in Hong Kong to another Central Business District is a good example of placemaking. Multi-governmental departments and businesses work with citizen to revitalise areas for economic growth and global competiveness, which also includes including liveability.

11.3.3   Sustainable housing (economics)

Public housing forms an important role in many Asian cities on meeting the ever-increasing demands on housing due the rapid urbanisation. Two typical examples of this are Hong Kong and Singapore, which both provide 30% and 90% of its household population. Their quest for sustainable housing plays a significant role of the city as whole on archiving the sustainability objective.

The Hong Kong Housing Authority (HKHA) is a statuary body responsible for developing and implementing public housing programmes to meet the housing needs of families who cannot afford private housing. They provide affordable subsidized rental housing for about 30% of Hong Kong's household population. HKHA owns a stock of about 700,000 flats, and are building an average of approximately 15,000 flat per year. The government has a mission of providing a green and healthy living environment for Hong Kong's community.[160]

In 1999, HKHA's Environmental Policy was launched to promote healthy living, a green environment and sustainable development while providing public housing and related services. In 2001, HKHA was the first to implement site planning and the design process of high-rise public housing with the application of microclimate studies to improve the environmental performance of its housing developments. The SARS outbreak in 2003 aroused public concern for healthy living. From 2004 onwards, all new public housing building projects applied micro-climate studies, and more than 30 public housing projects improved their new estate designs as a result of these studies. The application of comprehensive microclimate studies facilitates better planning and a higher level of sustainable public housing design that balances social, economic and environmental needs (Example 11.1).

In Singapore, the Housing and Development Board (HDB) is Singapore's public housing authority. Originally set up as a government agency to house the homeless, its focus these days is to provide affordable and attractive home ownership options that are intended to maintain Singapore's economic, social, and political stability.[161]

HDB flat is an important component of Singapore's identity. HDB provides housing to 80% of Singapore's residents, and 80% of HDB habitant households own their home. More than 1 million flats have been completed, with another 70,000 to be constructed annually, depending on demand.[162]

Due to the agenda of building affordable and quality homes for the majority of the population, HDB building form differs from public housing in many other countries. It is also the default Singaporean housing typology and new town planning and sustainability is integrated on a few different levels.

Socially, HDB estates feature community spaces for multi-racial and diverse communities to mingle and build coherence. Besides high-rise residential blocks, each township is supported by commercial facilities, transport nodes, sports centres, parks, places of worship, schools, libraries, and other community facilities. Every few blocks have facilities for the three-generations of a family. They feature a precinct pavilion, childcare centre, seniors' centre, and playgrounds for the old and young.

Economically, older housing estates are maintained and upgraded to ensure their vitality and to prolong their lifespan. Renewal programmes include the Main and Interim Upgrading Programmes, the Selective En bloc Redevelopment Scheme, and Lift Upgrading Programme. This reduces the need for investing additional resources in urban sprawl, which comes with a much greater environmental impact.

Recently, integrating new technology such as sustainable and smart initiatives has been a big part of new township developments on green field sites to ensure its relevance. For example, environmentally friendly features such as water sensitive urban design, pneumatic waste systems, dual refuse chutes, solar photovoltaics and vertical greening can be found in Punggol New Town. The "Smart HDB Town Framework" incorporates elements of smart planning, smart environment, smart estate and smart living. This will be piloted in the Punggol Northshore district (Example 11.2).

Example 11.1   Green designs of Upper Ngau Tau Kok
in Hong Kong, China

Upper Ngau Tau Kok (UNTK) is a public housing redevelopment project of the Hong Kong Housing Authority, comprising construction of 4,584 flats in six high-rise residential blocks for 12,200 people, a two-storey carport, a commercial centre, and a children's youth centre at podium level (Figure E11.1.1). To create a sustainable environment for residents, the project design underwent a thorough assessment on natural ventilation, daylight provision, solar heat gain, sunlight, wind microclimate, and pollution dispersion behaviour. The approach follows well with the framework shown in Figure E11.1.2.

The performance of domestic flats was studied in terms of natural ventilation, daylight provision, sunlight availability, and solar heat gain. Representative domestic flats at low (10/F), middle (20/F), and high level (30/F) satisfy the natural ventilation minimum statutory requirements of 1.5 air exchange. Overall, daylight performance for the studied flats is higher than the minimum statutory requirements of 8% vertical daylight factor for habitable rooms. Sufficient provisions of sunlight for drying racks reduce the use of gas or electric drying machines, and consequently energy consumption. Besides providing light and warmth, sunlight also contributes to human health. It helps the synthesis Vitamin D in the human body, kills germs, and keeps away a number of diseases. Lastly, the solar heat gain of a flat is determined by its orientation and possible shadowing effect from other buildings for the indoor thermal comfort of residents.

Figure E11.1.1   Upper Ngau Tau Kok Estate Redevelopment, Hong Kong. © Arup

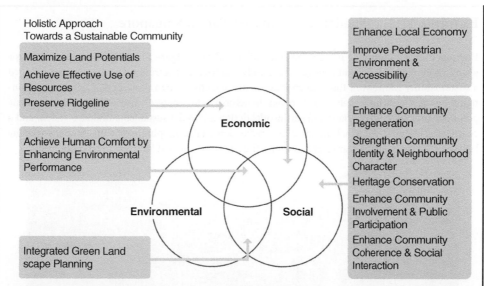

Figure E11.1.2 Sustainable Community Framework of Upper Ngau Tau Kok Estate, Hong Kong. © Arup

The natural ventilation and daylight performance at the main entrance lobbies of the six residential towers of the redevelopment were determined. For natural ventilation, the flow rates across the lobbies were found to be higher than the comfort benchmark. With mitigation measures by means of addition of a wing wall and higher louver location, the ventilation performance at the lift lobbies improved by at least 100%. For natural daylight performance, the daylight distribution of the main entrance lobbies under overcast sky conditions was studied. The glass block wall and the glazing located at the facade of the main entrance lobbies allow daylight penetration. Those areas near the glass block wall or glazing have higher lux levels, such that artificial lighting could be saved during the daytime. For those areas far from the glass block wall/glazing or deep inside the lobbies such as the lift waiting area, the daylight levels are below 250 lux and artificial lighting should be provided during overcast sky conditions.

For the microclimate analysis, the wind microclimate and thermal comfort at four areas including the Central Plaza, Pedestrian Walkway, and Basketball Court were investigated. For the Central Plaza, activities of sitting or resting were assumed. Sheltering is provided for the major pedestrian walkway, the overall thermal comfort is enhanced as compared with a situation without sheltering. A canopy is introduced with a slope downwards along the performance stage to the audience at Central Plaza. This arrangement brings the summer wind into the Central Plaza and enhances the overall wind microclimatic conditions. The basketball court provides a comfortable space for walking or running activities in summer and winter, as it is sufficient with summer winds (around 1-2 m/s) but silent with winter winds (less than 0.5 m/s on average). In conclusion, shading/sheltering structures are suggested to be provided mainly at the resting places at the Central Plaza and along the major pedestrian routes. Not only does the canopy enhance the air movement at the ground level in summer, but also provides partial shading for users, and thus reduces the mean radiant temperature that indirectly affects people's thermal sensation.

Lastly, the dispersal behaviour of the odour gas from both the commercial kitchen and Refuse Collection Points (RCPs) was investigated. The pollutant concentrations received at the nearby sensitive receptors were found to be below the benchmark pollutant level value.

Example 11.2   Bidadari estate masterplan in Singapore

Envisaged as "a community in a garden", the Bidadari Estate (Figure E11.2.1) will be an urban oasis where residents can relax and connect with family and friends in a tranquil setting. Arup developed a sustainability framework that incorporated water sensitive urban design (WSUD) and environmental modelling which enabled us to review benchmarks for energy, water and waste, and estimate potential savings. The drainage and earthworks design allowed a sustainable drainage system and a centralised attenuation, which doubled as a recreational lake, keeping intact much of the existing landscape. Locating the water reservoir under the park also allowed land use to be optimised.

Figure E11.2.1   Masterplan for Bidadari Estate, Singapore. © Housing & Development Board Singapore

11.4   NEIGHBOURHOOD ASSESSMENT

It is recognised that achieving comprehensive sustainability in a community is not quite possible when we work with one building at a time. Although practical designs and operations are found within an individual building, many constraints have normally been inherited already, when a neighbour is formed. Hence, a master plan of a district already exists. Therefore, similar to the green building certification rating mentioned in Chapter 7, the industry realised the need to create another tool that could guide the master planning of a neighbourhood with sustainable focus.

11.4.1   History of overseas schemes

LEED-ND was the first of its sort in the market with its Draft version introduced in 2005, with two subsequent versions over the following four years, during which the introduction of three other schemes – BREEAM for Communities,

Table 11.2   Overview of Five Overseas Schemes

	Year of Issue	Origin of System	Implemented Organisation	Nature of organisation
LEED-ND	2005	US	US Green Building Council (USGBC)	Private
CABSEE Urban Development	2005	Japan	Institute for Building Environment and Energy Conservation (IBEC)	Private
BREEAM Communities	2008	UK	Building Research Establishment (BRE)	Private
Green Mark District	2009	Singapore	Building and Construction Authority (BCA)	Government
Green Star Communities	2012	Australia	Green Building Council of Australia (GBCA)	Private

Green Mark for Districts and CASBEE for Urban Development – also occurred. The newest in the market is Green Star Communities, whose Pilot version was introduced in 2012 (Table 11.2).

11.4.2   Definition of community/neighbourhood

The literature review in terms of "neighbourhood" and "community" has found that the term neighbourhood is more spatially bound with physical extent,[163,164] whereas community could be either actual or virtual with much emphasis on connection and communication between critical points/programmes/places.[165]

The distinction between the two is ever stark when one talks about his/her own neighbourhood – close vicinity where he/she resides – as opposed to "online" or in-person communities, in which common interests and/or values bind people together (Figure 11.2).

As concluded by GBCA, there are a range of development types, building, and land use typologies that influence the language in describing a project. Precincts, neighbourhoods, and villages are all common terms used to describe developments that constitute a number of buildings and amenities.[166]

None of the five overseas rating schemes explicitly define the terms "community" or "neighbourhood development" for the following reasons:[167]

- The term "community" is a vague term that can be interpreted in many different ways with little consistency or consensus within research, development, and professional services industries;
- There is no real value in trying to establish a definition for community as its openness is potentially its strongest asset;
- The definition (for neighbourhood) has a "time" aspect - With improved accessibility and people's changing demands in today's society, existing definition, aspects and what constitutes a traditional neighbourhood may not be applicable to a contemporary neighbourhood.

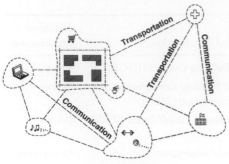
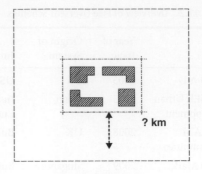

Communities	Neighbourhood
• Common locality, either in-person or online	• Spatial scale place-based community
• Interpersonal relationships of sociability, support and information, either in-person or online	• Communities may be based on a neighbourhood
	• Relates to the area around a residence within which people engage in neighbouring, which is usually viewed as a set of informal, face-to-face interactions based on residential proximity
• Common values, norms and interests, without necessarily being co-located	
• Place-to-place community, with little interaction with the intervening territory between places	• An area where the majority of people know the significant buildings and central focus of the area by sight most of those who live there
• Efficient transportation and communications	

Figure 11.2   Difference between Communities and Neighbourhood. © Arup

Hence, Green Star, LEED-ND and BREEAM Communities all express the view that these terms are best defined via their proposed prerequisites and credit criteria, which provide a scope for the developments and promote best practices in new community/neighbourhood developments today.

11.4.3   Assessment aspects/categories

There is a large disparity amongst the contexts of schemes that are designed to be assessed. The criteria for assessments are different. The table below shows a list of core assessment aspects as per each overseas scheme (Table 11.3).

For any project to qualify for a ratings scheme, one must demonstrate how a project satisfies all prerequisites of the concerned scheme.

LEED-ND has the most extensive list of prerequisites compared to other overseas schemes. Both LEED-ND and Green Star - Communities Pilot have adopted a balanced approach for all categories, except innovation, commences with prerequisites. While BREEAM Communities focuses on the pre-design stage of setting principle, Green Mark District emphasises green building practice and environmental planning. CASBEE focuses on the governance issues of project proponents, as projects under CABSEE usually consist of multiple participants and stakeholders including a form of joint venture amongst a number of developers.

Varying focus of each scheme's prerequisites can be interpreted as the scheme's intentions for targeted use, users and stages of projects for implementation. The prerequisites focus areas of each overseas scheme are shown in Table 11.4.

Table 11.3   Assessment Aspect or Criteria of Five Overseas Schemes

LEED- ND	BREEAM-Communities	Green Star-Communities	CASBEE- Urban development	Green Mark - District
• Smart location and linkage • Neighbourhood pattern and design • Green infrastructure and building • Innovation in design or innovation in operations • Regional priority	• Governance • Social and economic wellbeing • Resources and energy • Land use and ecology • Transport and movement • Innovation	• Governance • Design • Liveability • Economic prosperity • Environment • Innovation	• Natural Environment (microclimates and ecosystems) • Services Functions for the Designated area • Contribution to the Local community • Environmental impact on Microclimate, facade and landscape • Social infrastructure • Management of the Local Environment	• Energy Efficiency • Water Management • Material & Waste Management • Environmental Planning • Green Buildings and Green Transport • Community and Innovation

Table 11.4   Prerequisites Focus Area of Five Overseas Schemes

	Focus Area of Perquisites
LEED-Neighbourhood Development	Balanced and detailed approach - Prerequisite applies to each category (except innovation and regional priority) for projects seeking all award stages. Particular emphasis on promoting connectivity, efficiency in transport, building energy and water. The only overseas tool that establishes prerequisites on neighbourhood pattern and design.
BREEAM-Communities	Emphasis on establishing the principle of development (to improve sustainability that necessitates a site-wide response is understood and a holistic strategy for the site is encouraged.) Particular focus on consultation and engagement. It is also the only overseas tool that includes this aspect as prerequisites.
Green Star-Communities PILOT	Projects must be mixed-use development with clear definition of district. Initial project certification must be achieved within three years of registration A balanced approach – minimum points must be achieved respectively in each category (except innovation), in addition to achieving a minimum total score equivalent to a Four Star Green Star rating.
CASBEE- Urban development	If multiple entities of project participants are involved, the existence of a consistent intent for the project shared between the multiple involved entities is a precondition requirement. The entities must also form a common perception for the implementation and result of the assessment.
Green Mark - District	Promote the utilisation of green building practice. Specific energy efficiency requirement for District Cooling System. Emphasis on environmental planning, followed by sustainable construction and product use, for infrastructure and public amenities, for projects seeking to achieve the top two ratings (Green Mark Platinum and Gold Plus)

Four of the five overseas schemes studied (up to early 2015), listed in Table 11.4, with the exception of CASBEE for Urban Development, cater for international application outside its author nation. Perhaps due to its short time period since its introduction, Green Star Communities has not yet been able to attract any projects outside its home country of Australia.

Once again, LEED-ND is the front-runner with 27% of its registered and/or certified projects being located outside of the US. With the sheer number of projects currently taking place in China, the success of LEED-ND in this region is expected to continue for some time.

LEED-ND's success in international application is credited to a combination of the following reasons, which are presented in no particular order:

- Economic success and continual development in emerging markets, in particular, the Greater China region.
- Brand recognition, established by other LEED Schemes such as New Construction and Core & Shell, and so on;
- Positive accommodation and extensive support mechanisms to answer queries;
- Aggressive marketing.

Neighbourhood development is the core of creating a sustainable community. LEED-Neighbourhood Development (LEED-ND) has adapted the ten principles of "Smart Growth" and developed it into an assessment scheme. It has now become possible to incorporate the abstract concept into the design. The first LEED-ND certified commercial development in China was the Wuhan Tiandi, which was certified as Gold label in 2007 for its approved plan.

The key design principles adopted include mix-used development, enhanced local employment, walkable community, improved public transportation, and place making. It took more than 10 years for the entire 50ha site providing a total GFA of 1 million m^2 to complete the construction. Wuhan Yongqing Mixed Use Development (WHYQ) has the advantages of vibrant developments, proximity to downtown, intersecting transit connections, availability of undeveloped land, and people that make it an important strategic area for future urban development in Wuhan. Shui On Land initiated this study to examine the approved Master Layout Plan on the potential of an integration of sustainable urban design strategies, green infrastructure opportunities, innovative building systems and LEED ND label to derive a sustainable community design model (Example 11.3).

11.5   DEVELOPMENT OF BEAM PLUS NEIGHBOURHOOD IN HONG KONG

Neighbourhood development is the core of creating a sustainable community. In 2013, the Hong Kong Green Building Council (HKGBC) initiated the development of a rating tool to assess and promote sustainable development on a community/neighbourhood scale in Hong Kong (Figure 11.3).

Example 11.3   First LEED-ND project in China

Ten years on, Yongqing district of Wuhan has developed a sustainable community in which individual properties and the public realm function together as an environmentally low-impact unit with high economic potential and social benefits (Figure E11.3.1). This reduced impact and enhanced benefits has created significant value for each participant including developers, property owners, the community, the local environment, and economy. In particular, the developers and property owners can benefit from lower operating costs and increased property valuation through the creation of a strong neighbourhood identity. The environment can be improved by dramatically reducing the impact of urban development on local and global ecosystems. The community and its government can benefit through increased property valuation and possibly resultant taxes, reduced demand for future infrastructure investments, and an increase in future resource capacity generated for others outside the development. Together, these benefits allow for more secure long-range economic and community development with an environmentally and economically sustainable model in Wuhan and in China.

The adoption of the LEED-ND design helps the sustainable designs of individual buildings. Gradually, all commercial sites will adopt the LEED-NC design principles and certified. All the completed commercial buildings are LEED-NC certified at Gold level.

Implementing the LEED-ND criteria has faced many challenges in China. To facilitate all parties on the requirements, a Sustainable Community Design & Implementation Guidelines (SCG) was prepared. In the past decade, it has guided design teams and construction teams to implement the identified sustainable strategies at master planning, building design, construction, and operation stages. The SCG depicts the detail requirements at different areas of sustainable design as well as providing guidance on LEED-ND certification. The document should form part of the legal obligations that all parcel/building development shall be committed. All parcel/building developments must agree to follow the SCG requirements as a part of the development contract. The methodology for making these guidelines a statutory instrument must be agreed. The design quality, technical performance, and compliance with the SCG of each aspect of every development at building or site wide levels shall be reviewed and controlled via a transparent and auditable process led by experts.

Figure E11.3.1   LEED-ND for Yongqing District, Wuhan, China. © Shui On

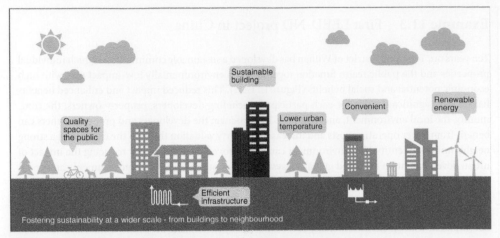

Figure 11.3   Concept Development for BEAM Plus Neighbourhood. © Arup

Based on BEAM Plus system for individual building, the key objective of the neighbourhood system is to recognise and encourage the positive impact in terms of environmental, social, and economical aspects of a project to on-site occupants and neighbours. The focus areas are the following:

• External spaces including major open spaces
• Community involvement and local character
• Synergy between buildings

The deficiency in building certification systems is the limitation of their applications in the public realm and spaces beyond one's own building/site. This also includes a lack of provisions to assess the collective performance of a cluster of buildings that make up a neighbourhood, community, district, or precinct.

A separate mechanism is therefore considered necessary to help fill the gap, especially at master a planning stage. A development is conceived and brought together to form a development of significant scale, intensity, and/or complexity with project implementation likely to be spread over a relatively long period of time.

11.5.1   Landscape and ecology in built environment

Hong Kong is a small city with 1,100 km² of total land area, of which only approximately 20% is used for development purposes, whereas the remaining 80% has restricted areas with natural reserves, country parks, and so on. Despite a low percentage of developed area in comparison to the total landmass, due to the concentration of urban developments, Hong Kong's urban areas are highly congested with high-rise buildings accommodating offices, commercial and residential spaces, roads, and other infrastructure necessary for the city's 7.1 million people.

As a rainforest before human arrived, Hong Kong has a subtropical climate with noticeable contrast in temperature, rainfall and relative humidity between summer and winter. Hong Kong's landscape displays various habitats and species, including much of which that are not native to Hong Kong. The altering landscape is often due to its location, wind environment, and other microclimate conditions that govern various types of vegetation in the area.

The main landscape and ecology issues, and the ways in which they are factored into each of the assessment tools, are described briefly below. There are also some discussion on the appropriateness and applicability of similar considerations being applied in the context of Hong Kong.

The issues may be broadly subdivided into those related to:

- Preservation and enhancement of existing resources;
- Provision of facilities for public use;
- Adoption of sustainable design details/features/methods.

11.5.2   Stakeholder engagement

There is a wide spectrum of aspirations and opinions from different stakeholders on the sustainability of the built environment and neighbourhood in Hong Kong. A stakeholder engagement exercise provides opportunities to gather stakeholders' insights from different fields on how a sustainable neighbourhood can be built. With the view of establishing a consensus, efforts on promoting sustainable neighbourhoods could be encouraged and merited through the proposed neighbourhood scheme.

With the involvement of various stakeholders in relation to the building industry, their aspirations and opinions could be gauged and analysed for incorporating the proposed neighbourhood scheme. This open and interactive stakeholder engagement process contributes to establishing support and ownership from different stakeholders on the recognition and implementation of the proposed Version 1 tool in Hong Kong.

Different stakeholder engagement activities were designed to understand views of stakeholders on the following:

- Understand the practicality and relevant concerns in adopting sustainable design and practices in the proposed Version 1 tool in the context of Hong Kong;
- Solicit their views and recommendations on implementing the proposed Version 1 tool to assess their projects;
- Explore ways to encourage the adoption of sustainable design and practices, with a view to facilitate acceptance from the industry.

Stakeholder engagement activities were carried out after the formulation of the draft Version 1 Tool, in order to better understand the views and aspirations of the industry and the professionals. Representatives from the following groups (Figure 11.4) were invited to take part in the multi-stakeholder engagement exercise. These include:

- Government bureaux and departments;
- Professionals;

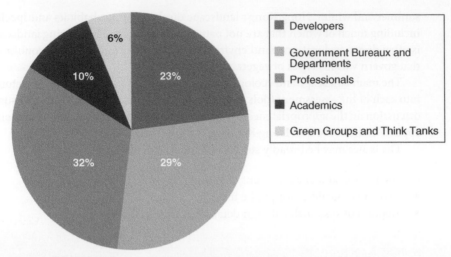

Figure 11.4   Distribution of Pre-Consultation Questionnaire Respondents. © Arup

Figure 11.5   Public Consultation Meetings in Process Developing Scheme. © Arup

- Academics;
- Green groups and think tanks;
- Developers (including private developers, quasi-public organisations engaging in estate development university campus development offices, Theme Parks and NGOs with property development potential, etc.);
- Others.

General support for the proposed Version 1 tool was received in the stakeholder engagement forums (Figure 11.5), attended by stakeholders from various backgrounds, including the government, quasi-public organisations, professionals, academics, green groups, think tanks, and developers.

In the stakeholder engagement exercise, the interim rating scheme (BEAM Plus Neighbourhood Version 1) was discussed with different stakeholders. As demonstrated in the previous sections, most of the comments from the stakeholders were incorporated in the revised neighbourhood scheme.

The multi-stakeholder consultation forums provided an ample opportunity for a better understanding of the local needs and the further refinement of the all-rounded tool that can easily be accepted and adopted by the industry. It reminded the Consultant of growing needs and specific requirements of sustainable neighbourhood development in Hong Kong.

Various stakeholders offered their experience in the industry and how the tool may help to proliferate a better design and development practice in highly company and highly dense environments of Hong Kong. Given the variety of participants' backgrounds, there were colourful and sometimes conflicting opinions and suggestions to take into consideration. Some examples of conflicting comments include, tightly controlling or completely omitting any exclusions where as some advocated for increasing exclusions on building-specific credits, to cater for existing neighbourhood projects. Such contrast was most obvious during the discussion on the scale of potential projects, which in the end, was concluded that there was no need to specify a specific range in size given the complex and high-density built environment of Hong Kong. It was also concluded that not having a limit would not impede the tool in achieving its objectives of advocating sustainable neighbourhood development through benchmarking.

11.5.3 The establishment of BEAM plus neighbourhood

The BEAM Plus Neighbourhood Version 1 tool has been refined to address the contexts of Hong Kong as follows:

- Social - Depending on location, geography, existing and proposed land uses and infrastructure, each neighbourhood has its own needs, challenges and opportunities. The tool promotes neighbourhoods to adopt a place-making approach to build on their unique strengths and local character. The tool promotes project proponents to strive for sustainable development in the community engagement for design brief formulation, consideration for social equity in establishing development mix, sharing of neighbourhood amenities and resources, integration with surrounding areas, conservation of cultural heritage, and existing uses.
- Economic - The tool promotes a holistic approach in the consideration of economic gains and sustainable operation of neighbourhood developments. A mixed use for a certain degree of self-sufficiency with no net loss of employment is promoted by the proposed tool. From a life-cycle perspective, provisions and the sharing of neighbourhood amenities, district infrastructure with surplus to cater for sustainable developments, creation of places is encouraged.
- Environmental - This tool places a great deal of emphasis on benchmarking environmental quality and performance of neighbourhood development with particular considerations for high-rise and dense developments, and their impacts/contribution to micro-climate, human comfort and wellbeing of the public and occupants using external spaces, ecological values of land, waste and water management. Integrated design to address continuity between private and public spaces, both within and beyond the site boundary, is promoted by the tool.

11.6   SUMMARY

The sustainability building design and operation framework has been widely adopted worldwide, and most locations have their own specific systems tailor-made for various places. However, building-oriented framework has limitations in the control of the public realm and spaces beyond one's own building, and it is hard to make up a neighbourhood or community. Hence, people have been working on the sustainability framework for a community to better cover the site scale aspects including social well-being, transport, neighbourhood design, and economy, and so on.

Chapter 12
Low carbon living

12.1 INTRODUCTION

The current patterns of production and consumption in cities are not sustainable. In developed countries, people use natural resources that require seven planet earths to be sustainable. In Asia, the situation is not much better. Citizens of Japan, Singapore or Hong Kong consume resources that require the equivalent of three planet earths. Changing our way of living is a prerequisite for a sustainable future throughout the world.[168,169] To sustain the behavioural changes, we need the design, not only hardware, but also software that can inform or educate people to thereby influence occupants of buildings to act responsibly.[170,171,172] The process of creating such "change agents" requires an understanding of the cultural contexts and traditions of local people. Cities in Asia are experimenting with their tactics, and many successful cases have been found to be valuable for other cities.

Four cases are presented: starting with education and public involvement of sustainable urban living in high-density Hong Kong, and moving to a demonstration of eco-experience of zero energy living in South Korea, then illustrating the possibility for urban farming in Japan, and finally creating LOHAS living in a resort/vacation facility for promoting a change to a sustainable eco-friendly lifestyle in Taiwan. These cases involve educating the general public on the importance of low carbon living and going from small to large in terms of scale, to let the people experience this lifestyle. The projects were thoughtfully planned and a lot of state-of-the-art technologies and systems were incorporated into the design to encourage behavioural change.

12.2 CARBON FOOTPRINT OF URBAN LIVING

As money measures cost for activities in economic terms, carbon footprint is the equivalent in environmental terms, based on the implications associated with living in cities. Everyone has a carbon footprint: greenhouse gases are produced when one uses electrical appliances in a household, for example, washing machine,

Building Sustainability in East Asia: Policy, Design, and People, First Edition. Vincent S Cheng and Jimmy C Tong.
© 2017 John Wiley & Sons Ltd. Published 2017 by John Wiley & Sons Ltd.

refrigerator, and so on, or when one travels by car, ship or airplane, or when visits a cinema, or goes shopping. All these activities constitute to a person's "Carbon Footprint", which is a measure of the amount of greenhouse gases produced by one's activities. A carbon calculator estimates the size of your carbon footprint and characterises your carbon emissions in daily life: at home, work, driving, commuting, education, entertainment, shopping and consumption, holidays, flying, and so on.

In 2011, as seen in Figure 12.1, Hong Kong's CO_2 emission per capita was 5.7, which only took into account emissions in the city and did not include emissions associated with imports and travel.[173]

Shared by a large population with a developing status, in China and India, the CO_2 emission per capita are as low as 6.7 and 1.7 tonnes per year, respectively. Owing to economic activities and lifestyle, CO_2 emission per capita is relatively low. However, as discussed in previous chapters, the two countries are rapidly emerging and developing into the most populous countries in the world. A chain of effects from urbanisation will likely follow, and eventually CO_2 emissions per capital will increase accordingly.

In Japan and South Korea, CO_2 emission per capita is 9.3 and 11.8 tonnes per year respectively, whereas Singapore is has reduced to as low as 4.3 tonnes per year. They are countries that have industrial activities started as early as the 1970s. Living standards have been improved and people are better off compared to other surrounding Asian countries. Island countries also promote the import and export of more goods and travel by sea and air. Their economic activities and lifestyle have produced higher CO_2 emissions per capita.

Low carbon living is about creating a lifestyle that produces low/less carbon emissions. There are always ways to reduce carbon emissions (carbon footprint) in daily activities, that is, at home, school, work, and leisure. Simply switching off a light can effectively reduce carbon emissions. Some pieces of advice include:

- Changing your light bulbs
- Planting trees
- Buying energy efficient appliances

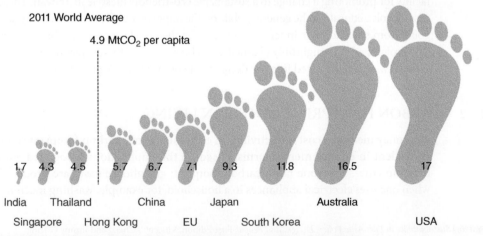

Figure 12.1 Carbon Footprint per Capital for Selected Countries in 2011. © Arup

- Conserving energy
- Using water efficiently (hot and cold water)
- Buying clean energy
- Driving less—driving smart
- Eating green
- Reducing waste—recycling
- Offseting your emissions

12.3   BEHAVIOURAL CHANGES

The behaviour of individuals can have an immense effect on the overall sustainability of a society. Although many people are concerned about the environment, this does not always translate into taking practical steps to reduce their environmental impact, that is, sustainable behaviour. Most of the behavioural change programmes in communities are trying to influence individual behaviour via economic instruments such as grants and rebates, or via education and persuasion.

This type of behaviour is exemplified on the mode of consumption (i.e., green purchase) on the way we use our appliances everyday (e.g., switching unnecessary lifts off) or collectively forming a green community (i.e., through social media). EU experience has suggested that behaviour change is the most cost-effective instrument to achieve energy saving targets, and probably the last method if other technical means have been exhausted.

The behaviour of people on the use of building is complicated. Certain behaviour can be affected by psychological, social and environmental factors. Studies[174] were conducted by IEA on understanding the dominant factors that will help behavioural changes. Figure 12.2 shows the process of how the occupants' behaviour impacts the building's energy usage. Occupant behaviour is driven by different motivations, such as thermal comfort that is considered as personal and cognitive factors, group norms and organisational culture that are environmental factors, and habits as behavioural patterns.

Before thinking of the methods to encourage changes, it is important to identify the ways that individual would react to situations where people can adapt and adjust to a built environment. When people experience changes in the environment, the body will show signs such as changing pupil size, ear sensitivity,

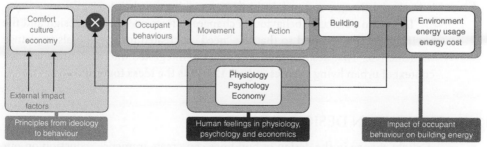

Figure 12.2   Behaviour Model of IEA Annex 66: Definition and Simulation of Occupant Behaviour in Buildings. © Arup

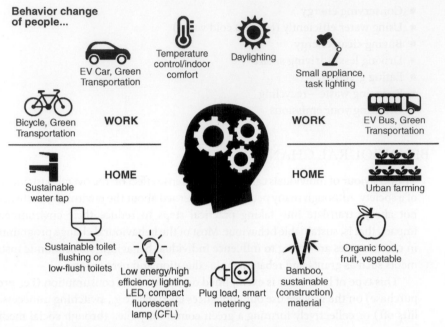

Figure 12.3   Embedding Behavioural Changes in Life and Work. © Arup

and sweating. People can make individual choices on wearing glasses and earphones, and change clothing level and material. They can also make adjustments of their surroundings by switching on or off the lighting, opening or closing windows and doors, and turning on or off the ventilation systems. They can even move away from the current situation by changing the position or escaping the location all together. All of these options are essential to recognise since they can help the designers to incorporate methods that can determine the success of implementing the behavioural changes.

The drive for "Behaviour Change" dominates public disclosure on sustainability. Design is implicated by supplying "sustainable" products intended to covertly influence users to enact more sustainable behaviours – such as saving water or energy – or by supplying "educational" messages about what people should be doing differently. More often than not, sustainable designs are not practiced.[175] Others are proposing using technology, in particular the recent advancement in ICT for public display for behavioural changes and sustainable living, making a real-time city.[176]

To sustain particular behavioural changes, efforts on building design and the environment are required so that the "need" is created for individuals to change their lifestyle. That design has to be able to be integrated into the culture and the context of urban living in cities. Figure 12.3 puts the ideas together.

12.4   CHANGES IN DESIGN CULTURE

A slight change in the design of buildings can create immense reduction on our carbon footprint. Many of these designs will have little impact on our practices, but will change our habits for energy and commitment to environment. Chapter 9 introduced the strategies on reducing energy use in buildings. More savings can be generated if we look into the problem from the perspective of occupants. The following strategies can change the behaviour of occupants.

12.4.1   Task lighting

Energy use in buildings for providing lighting accounts for as high as 25% for a typical office building. In the past, savings in lighting energy were relatively simple from the perspective of design, involved only of selecting high efficacy light fixtures and the associated control. A new practice of task lighting is beginning to prevail in the market. It has the potential of reducing lighting energy by half. In the past, a lighting power density (LPD) energy use indicator of lighting was designed at 20 W/m² to deliver a 500-lux lighting level. Nowadays, the LDP can be brought down to 8 W/m² with the provision of task lighting. The reduction in energy use for task lighting is significant.

12.4.2   Thermal comfort

Space cooling is essential for offices for Asian city climates. It requires a lot of energy to provide a thermally comfortable space in Asia due to the climate.[177] There is a culture in the practice to design the set-point temperature to a very low temperature. In operation, some offices maintain indoor temperatures as low as 21°C, which is well below the design value of 24°C. This creates a huge wastage of energy. In fact, one degree higher of set-point temperature can save 2% of HVAC energy. In many Asian countries, the governments are advocating a high set-point temperature. In Hong Kong, it is 25.5°C, whereas it is 26°C in China, and 28°C for Japan.

  As for heating, it is a fact that if the temperature is set too high during winter, the relative humidity will drop to too low for comfort. Instead of central heating, spot heating will be more effective on keeping individuals at a comfortable level. Therefore, it requires a fundamental change of our culture on our perception of "comfort" before we can see any significant step change.

12.4.3   Natural ventilation

Before air conditioning, natural ventilation was used in all cultures. Since it is possible to have control of setting a comfortable spatial temperature, modern buildings have been adopting mechanical ventilation by means of using energy. Natural ventilation is making a comeback in sustainable building designs in order to achieve energy reduction. Incorporating natural ventilation into a building is not straightforward. It needs to be intentional as the building orientation first determines the feasibility, then it will take a detailed design of window configurations, both size and arrangement, to allow it to be effective. However, a building can be designed for using natural ventilation, but occupants might not choose to utilise it for many reasons. Therefore, as mentioned previously, it will take education and change of culture for maximising the energy saving from natural ventilation.

12.4.4 Green products

Green procurement is now a very effective means of reducing energy use (in UK, the Green Deal) for households or commercial use. Many countries have energy labelling schemes, which are voluntary or mandatory schemes. It is now a practice in the industry to procure green products, not limited to energy saving products but also environmentally friendly materials to improve the overall sustainability of buildings.

12.4.5 Smart metering

Smart meters aim to increase the prominence of energy use at home and office by providing feedback on energy use. Smart meters enable accurate energy billing and real-time measurement of electricity use and can communicate this information through a customer display. Smart meters could change people's habitual energy use in two ways: reducing overall energy consumption and shifting energy consumption.

12.5 ECO-EDUCATION

Behavioural change often is not so easy and not quite possible to change in a short period of time. Often, it is required to develop an easy-to-understand knowledge and implementation plan so that they can be conveyed to others. One important platform for this to happen is through education. Although education touches a very broad range of many aspects, and there are countless methodologies that can help to develop programs that would be more effective in achieving a desirable goal, only a few examples are included to illustrate how actual projects can show the general public what can be done to build a more sustainable community.

Lack of public awareness on the impact of carbon footprint is one of the reasons of high carbon living. A demonstration of a positive living attitude is needed in Hong Kong. This Eco Home, as shown in Example 12.1, is a combination of education and demonstration apartment to showcase the innovation and commitment of New World Development on developing an eco-friendly future of Hong Kong. By integrating indoor comfort, water conservation, natural materials, and a change of lifestyle, Eco Home will introduce truly sustainable living for the future. Furthermore, smart homes are a growing trend. With the increasingly adoption of the Internet of Things, homes can be equipped with sensors that collect data on, monitor, and manage energy usage (Example 12.1).

Example 12.1   Eco-home in Hong Kong, China

Demonstration of Future Home

One of the major intentions of Eco Home is to demonstrate a realistic definition of a future home (Figure E12.1.1). Although high-tech intelligent homes are the general public's perception of a future home, a truly sustainable future home should also achieve a better indoor environmental quality through technologies in various aspects, such as thermal sensation and visual comfort optimisation (e.g., bladeless ceiling fan, LED pixel ceiling, fibre optic lighting, electro-chromic glazing). The Eco Home will be fully equiped with sensors and smart control systems which can be linked to mobile devices for remote control use. The Eco-kitchen will provide a cradle-to-cradle solution to our daily domestic waste, which will greatly reduce the load on controversial land-fills in Hong Kong. Indoor hydroponic farming provides an innovative solution to low impact and LOHAS lifestyle. Home farming not only provides vegetables for consumption without the emissions transportation, but also acts as an indoor air purifier.

As a Centre of Education

Education is another major purpose in Eco Home. Aside from the display of state-of-the-art technologies and eco-friendly features, the eco home will allow visitors to raise awareness on typical home energy consumption by visualisation and to experience what they can do to reduce their carbon footprint (Figure E12.1.2). The educational experience is designed to be an interactive process rather than a still display. Visitors will be given a RFID wristband with a unique ID to be carried around during their stay in the Eco Home. Visitors can choose to enter their personal information which will then personalise their experience in the Eco Home. The wristband can communicate with each feature in the Eco Home and as the visitor walks through each section and interacts with the displays, all the interactions and scores will be stored in the system for viewing at the end of the tour. Moreover,

Figure E12.1.1   Rendering for Eco-Home, Hong Kong. Photo source: Zero Carbon Building

physical interaction is incorporated in this project, for example, electricity generating exercise equipment, electricity generating floor, as it promotes the eco-living awareness and achieves a better educational purpose (Figure E12.1.3).

The Concept of "De-Carbonization"

The average amount of annual carbon emissions of a Hong Kong resident is at 5.7 tonnes. According to a research paper titled "Carbon Footprint of Nations: A Global, Trade-linked Analysis", Hong Kong has the second highest carbon footprint per capita in the world emitting a stunning 29 tonnes per year. In ECO Home project, a brand new concept for demonstrating how one can live an eco-life called "De-Carbonization" will be introduced. As the visitors walk through the Eco-Home and experience each of the eco-features, the wristband records their choices and reports their total of carbon saving at the end of the exhitbition.

Experiencing

There are a total of four designated zones in the Eco-Home, each representing a user experience namely, Aquatic Equilibrium, Low Energy Comfort, Natural Materials, and LOHAS.

Aquatic equilibrium, as the name implies, is the concept of creating a closed-loop water system within the residential unit. In Eco-Home, the goal is to mimic the closed-loop water system with innovative low-flow fixtures, greywater recycling, and state-of-the-art washing machine.

Low-energy comfort is used to promote indoor environmental quality (IEQ) without the use of energy intensive systems (Figure E12.1.4). By adopting the natural resources into the design of the home, visitors can enjoy the comfort of nature such as natural daylight and natural ventilation, at the same time greatly reducing the carbon footprint of the Eco-Home.

The Leisure and Education zone is designated to educate visitors on sustainability through interactive games and activities. Through fun and interactive games, sustainability can be engraved deep into the mindset of future generations.

LOHAS zone is designated to promote self-sustaining life style and low impact living. Cradle-to-cradle solutions will be displayed to educate visitors on reusing household waste rather than

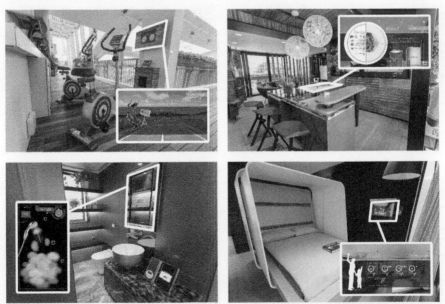

Figure E12.1.2   Monitoring and Educational Display at Eco-Home, Hong Kong. Photo source: Zero Carbon Building

dumping it into a landfill. In addition, visitors can participate and interact with the displays to allow them to experience LOHAS and learn how they can make a change.

In addition to the four designated zones, Eco-Home would also display the most eco-friendly building materials, furniture, and cleaning products in the market throughout the home to promote an environmentally-responsible source of materials and reduce respiratory or chronic diseases caused by irresponsible suppliers.

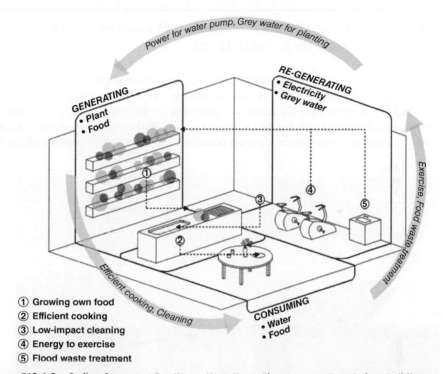

① **Growing own food**
② **Efficient cooking**
③ **Low-impact cleaning**
④ **Energy to exercise**
⑤ **Flood waste treatment**

Figure E12.1.3   Cycling Concept at Eco-Home, Hong Kong. Photo source: Zero Carbon Building

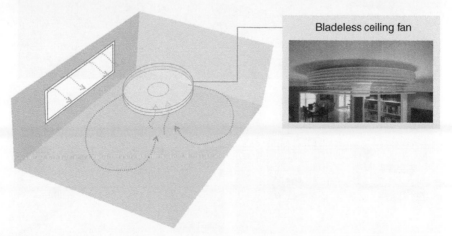

Figure E12.1.4   Ventilation Design at Eco-Home, Hong Kong. Photo source: Zero Carbon Building

12.6   ZERO ENERGY LIVING EXPERIENCE

Moving from a demonstration home to educate the public about possible ways to improve individual carbon footprint, it is possible to take another step forward to a zero energy house that integrates more sustainable technologies together. An example of a zero energy house is displayed in Example 12.2.

Example 12.2   Living a life the sustainable and smart way – green tomorrow in South Korea

Located in South Korea's Yangin city, Samsung's Green Tomorrow eco-development includes a 423 m² zero-energy house and a 298 m² public relations pavilion, as seen in Figure E12.2.1. Operated by Samsung C&T, it was designed as a sustainability showpiece for the entire country. Consultancy firm Arup provided the sustainable building designs as well as LEED consultation, working closely with Samoo Architects to create the best project possible. It was the first zero-energy building in

Figure E12.2.1   Green Tomorrow in South Korea. © Samoo Architects & Engineers/Samsung C&T

Photovoltaic Blinds

Direct Current Electric Power system

Standby Power Cut-off System

Emergency Fuel Cell

Figure E12.2.2   Features in Green Tomorrow, South Korea. © Samoo Architects & Engineers/Samsung C&T

Korea, and 68 state-of-the-art energy technologies were adopted. Green Tomorrow's energy usage has been reduced to 56% compared with an average common domestic household in the country, and has renewable energy, which is generated to power the remaining 44% of its required energy.

Green Tomorrow is not only an exhibition of future technologies, it has been called an "evolving building". Researchers and engineers continuously study and upgrade technologies. Finally, Samsung introduced a zero-energy residential building in 2013. Samsung Electronics took care of the heat, ventilation and air conditioning systems, and the control system to ensure optimum energy efficiency.

Green Tomorrow exhibits three key achievements: zero energy, zero emissions, and green IT. Arup's sustainability consultants investigated Green Tomorrow's sustainable features based on the categories including site, water, energy, atmosphere, materials and resources, and indoor environment quality.

Active design covers the membrane bio-reactor – an innovative biological treatment system. This allows grey and blackwater to be reclaimed for irrigation, toilet flushing and cleaning purposes. Together with water-saving equipment, this brings more than 70% worth of saving for potable water annually. Green Tomorrow collects domainant sustainable features and energy efficient technologies, such as the ones shown in Figure E12.2.2, and gives an educational demonstration for the community.

One of the exmaples of a zero energy house focuses on three major themes:

- Zero Energy - it optimises energy performance through reducing energy consumption and generating renewable energy on-site such that the generation amount is equal to or exceeds the consumption. The overall energy balance will be at net zero or plus.
- Zero Emissions - it eliminates CO_2 emissions during construction to operation, and to demolition by using sustainable materials and recycling waste.
- Green IT - it applies technologies that increase the efficiency of sustainable energy and creates a comfortable living space targeting the needs of people.

12.7   COMMUNITY CENTRE

With homes and houses that are able to show workable solutions in personal living, the next step is to continue to extend the sphere of influence to a larger context. In an urban setting, people are living closer and closer, not necessarily bringing their relationships closer, but certainty increasing the number of people in a same neighbourhood. Therefore, different types of community centres are required to provide various functions to serve the people.

There are four key concepts for a sustainable community centre to achieve:

1   Role model of sustainability - Sustainability is the key concept for a development, which becomes a role model for future developments for low carbon emission and a prototype for innovative sustainable technologies.
2   To demonstrate a coexistence model - Commercial activities and environmental protection are believed to be irreconcilable. The goal is to establish a complementary linkage, a demonstration that the coexistence of commercial activities and environmental protection is not simply a theory.
3   To create a focal point - A sustainable community centre forms a hub to promote sustainable technologies, community event space, and an organic market when it is within close proximity of agricultural land, vernacular villages, and country parks.
4   To showcase holistic and sustainable living - A sustainable community centre showcases a holistic and sustainable living style to visitors through interactive engagement and an implementation of sustainable design features.

Integrating architectural design, engineering and program, the development took a first step to shape a sustainable building in which environmental protection and commercial activities can coexist, which is demonstrated in Example 12.3.

Example 12.3   Green Atrium in Hong Kong, China

The Green Atrium, seen in Figure E12.3.1, is built within the residential development of Park Signature in Yuen Long, Hong Kong, where large amounts of traditional farmland is located. With this favourable environment, the project is designed to establish the coexistence of commercial activities and environmental protection, which are often irreconcilable. The environment provides the building with energy and commodity for commercial activities through the sustainable features in the upper block, whereas the commercial activities in the lower block provide resources for environmental protection work.

The sustainability objectives of the development are to embrace LOHAS, to construct a low carbon building, to create a sense of community, to educate the public for behavioural change, and to demonstrate the possibility of living synergy. These objectives cover the areas on energy, water, waste, air, and food. For environmental practice, a total 28 sustainable features are installed, which contribute not only to energy reduction but also the lifetime of other resources such as food and water, which can be maximised by creating consumption and waste loops.

Architecture design itself is integrated with sustainability consideration. The building embraces comfortable semi-outdoor space by the use of an innovative ventilation system, Air Induction Unit (AIU), as seen in Figure E12.3.2. Hong Kong is a sub-tropical city that experiences humid and hot weather for over half of any given year. During that time, the provision of thermally

Figure E12.3.1   The Green Atrium at Park Signature, Hong Kong. © Arup

Figure E12.3.2   AIU Application at The Green Atrium, Hong Kong. © Arup

comfortable spaces generally relies on air-conditioning. The excessive use of air-conditioning can lead to a significant increase in overall energy consumption and thus carbon dioxide emissions. Though smart design measures, thermal comfort can be achieved through alternative methods other than air-conditioning in a space. One of the smart design measures is to increase air velocity.

Other than conventional fans, AIU was invented to enhance air movement. The merits of the AIU could minimise the electricity consumption, whereas it could be easily incorporated with the architectural design with aesthetic appearance.

Sensors and metres measure energy and water usage during post-occupancy phase, renewable energy generation, and microclimates. This data is analysed and feedback is provided to improve ongoing operations. A performance dashboard is located at main entrance, which helps visitors to understand the building's performance. With the aforementioned combination of various passive design features and active devices, the building is anticipated to have 35% emission reduction, compared to the local standard, representing a reduction of 125 tonnes of CO_2 per year.

12.8   URBAN FARMING

Recently, the trend of urban farming is to integrate the farming system to the built environment, either within an outdoor or indoor environment. For example, the aquaponics system and hydroponic system can be integrated into the landscape and water feature systems, in order to be clearly visible and increase the participation of people (Figure 12.4).

Having completed in 2003, in the middle of Tokyo Metropolitan City, the podium roof garden of Roppongi Hills offers residents and workers an opportunity to experience farming and exposure to nature. Through rice and vegetable farming workshops, participants, including adults and kids, can deepen their understanding of the traditional food culture of Japan and the importance of the natural environment and food. The podium roof garden provides not only opportunities for such experiences but also offers a sense of community for the people, which is now increasingly sought after for people, especially in large and transient cities like Tokyo.

The podium roof garden is designed to work as part of a seismic damper in the event of an earthquake. Also, the thick layers of soil, greenery and water surface are expected to mitigate the Urban Heat Island effect and cooling load during summer.

Traditionally in Japan, maximising the use of resources with little waste is very well appreciated. The commercial market players such as property developers can brand themselves with a high social value by demonstrating sustainable operations and lifestyle, and as a result, they can potentially influence people's behaviour.

Figure 12.4   Concept of Landscape-Integrated Aquaponics System. © Arup

12.9   LIVING LOHAS

Lastly, promoting a Lifestyle of Health and Sustainability (LOHAS) is getting a culture to change and embrace low carbon living for a viable and desirable future. LOHAS focuses on living healthily and creating a sustainable eco-friendly economy. Personal choice on healthy organic food and exercising leads to a changing mind-set for purchasing eco-friendly products and services. Therefore, getting people to adopt LOHAS will help to build a low carbon future; one way to attract people to try out this new lifestyle is through a resort/vacation type of environment, where people are more relaxed and open to new ideas for a short period of time. Once they have experienced how easy and possible this lifestyle is, they are more likely to adopt it in their daily lives.

Huilanwan Sunrise Village is an example of LOHAS. This project consists of three 14-storey residential buildings, shopping mall, art village, and medical cosmetic clinics. The whole project adopts a hotel-style property management. The project emphasises that it is a place that people can stay for long or short period, or use it for vacation, health, or investment purposes. It has achieved the Leadership in Energy and Environmental Design (LEED) Gold certification for energy conservation. It is a shortlisted entrant of the MIPIM Architectural Review Future Projects Awards 2014 and the awardee of the 2014 Architizer A+ Award for the Best Architecture + Aging building.

Huilanwan Sunrise Village is far more than a stylish construction from the total design, from exterior, interior, to furnishing. It is an organic carrier providing smart living. Further information can be found in Example 12.4.

Example 12.4   Huilanwan sunrise village in Taiwan

Vision

The vision of design is to provide a framework for life to unfold. All residents share the same beliefs and are eager to exchange knowledge in the community, where we can create on the basis of lifestyle, health, and sustainability (LOHAS).

Health

A healthy lifestyle is becoming increasingly popular nowadays. In Huilanwan Sunrise Village (Figure E12.4.1), some organic gardens were inserted into green roofs and landscapes. All residents can harvest food that they planted, and visitors can also enjoy healthy foods in the restaurant (Figures E12.4.2 and E12.4.3). The nutritional information will be accurately displayed on packaging, menus or signage, which allows consumers to make informed dietary decisions. Fitness facilities such as a gymnasium, swimming pool, and an indoor jogging path (Figure E12.4.4) allows for a seamless integration of exercise and fitness into everyday life to support an active and healthy lifestyle.

Figure E12.4.1   Huilanwan Sunrise Village. © Taiwan Innovation Development Corporation \BIG

Figure E12.4.2   Organic Farm and Foods Support a Healthy Lifestyle. © Taiwan Innovation Development Corporation \BIG

Wellness

In the concept of LOHAS, the mental and physical health is inextricably connected. This development does not only provide healthy spaces or facilities but also actively organises learning clubs to satisfy residents both mentally and physically, which blends well in local contexts, such as Sailing Club or Flying Club. Diverse venues with nature and a pedestrian path winds between and around buildings like a hillside trail are designed to encourage social interactions throughout the site.

Sustainability

Sustainability is one of the major concepts of LOHAS. In this design, the building's orientation and landscape is conducted by computer simulations for a comfortable environment filled with daylight and natural ventilation (Figure E12.4.5). In summer, wind flows between buildings and is cooled by water features to improve thermal comfort. In winter, outdoor spaces are protected from cold airflow by buildings. Green roofs are introduced to reduce the urban heat island effect and decrease the solar radiation influence on the indoor space located at the top level. For an energy efficient and healthy indoor space, enhanced natural ventilation is offered to improve the indoor comfort level and reduce

Figure E12.4.3   Coffee Alley. © Taiwan Innovation Development Corporation \BIG

Figure E12.4.4   Jogging Path Winds Between and Around Buildings. © Taiwan Innovation Development Corporation \BIG

Figure E12.4.5   Computer Simulations. © Arup

Figure E12.4.6   Indoor Space is Filled with Daylight. © Taiwan Innovation Development Corporation \BIG

energy consumption. Other strategies such as PV, Rainwater Recycling System, BMS, Energy Efficient Lighting and A/C all respond well to the design concept of sustainability. It is expected to save 58,000 kWh energy, 60,000 tonnes of water, and 30,000 tonnes of CO_2 emissions annually, compared with similar building of this scale in Taiwan. All visitors can stay here for long or short periods with plenty of daylight, wind, nature, and enjoy the advantages of a sustainable design (Figure E12.4.6).

12.10   SUMMARY

The change in behaviour needs design deliberation to create the "change Agents". In Hong Kong, Taiwan, Korea, and Japan, cases were identified to influence the configuration of built environment designs and operations.

Section 5
Way forward

Chapter 13
Conclusions

ON CONTEXTS ...

The world is facing a challenge that is unprecedented in the history of mankind – climate change. People around the world are looking for solutions to solve this ever-increasing problem. Instead of focusing only on the Western developed countries, the hope should rest in Asia, where more people are living with human activity that is causing carbon emissions, will intensify due to economic development in the coming years.

Rapid urbanisation in Asia has created many economic miracles in the region. Asia is becoming the most important region for production of goods and doing business. More and more capital and talent are moving to mega-cities in the region. Asia has the potential to be the leader in sustainable development as more new buildings and infrastructure are being built to support economic development. Yet, many countries in region are facing social and economic changes that need the right form of governance and leadership to change the behaviour of the business as well as the individual. There is the need to not allow the region to become a consuming society.

The unprecedented urbanisation occurring in Asia also has many downsides. It has stretched the carrying capacity of the local environment to its limit. Extreme weather conditions caused by climate change pose a risk to people's life and property. Developing countries such as China are more vulnerable where the intensive process of urbanisation has resulted in severe environmental problems such as air quality affecting the health of people. How to build the cities' resilience to climate change and environmental crises is at the top of the agenda for many political leaders and policy makers in the region. Developed countries of the region are not better off. These countries are consuming more and more resources, energy in particular, to improve people's quality of live and sustain their reckless consumption behaviour. This results in a vicious cycle uncontrolled carbon emissions. Unless a sustainable development is in place to rectify the situation, more and more natural disasters and environmental catastrophes are to be expected.

Building Sustainability in East Asia: Policy, Design, and People, First Edition. Vincent S Cheng and Jimmy C Tong.
© 2017 John Wiley & Sons Ltd. Published 2017 by John Wiley & Sons Ltd.

Global and regional collaborations on implementing sustainable developments have been active in recent years. Discussions on governance, green business, and practice have intensified and from regional integration. A new mechanism of green movement is being formed in the region. It relies on strong interactions between government policies, practice in the market and the behavioural change of people to archive the objectives of sustainable development. This starts from policies, designs, people, and finally the market to transform the current practice and reverse the course (or the mode in specific) of the urbanisation process. The ecological age of the built environment can only happen through the joint efforts of governments, stakeholders, and advanced engineering approaches.

ON POLICY …

The right governance is pivotal to the success of implementing sustainable development. At a government level, a policy framework should be in place to direct stakeholders. At the top of the framework, policy priority should be formulated. There are many drivers (external and internal) that will influence this priority. Creating business or technology advantages for developed countries, whereas developing countries will have to focus on addressing environmental challenges created by urbanisation. Notwithstanding, creating an equitable society is prevailing policy objectives to ease the tension between the government and the people.

Regarding policy implementation, applying the right instruments by the government is the key to success. At the early stages of green movement, regulations should be in place to control business operations and individuals. When the green market is formed, the government can consider applying economic instruments, which has proved to be effective on incentivising the market. Some businesses, such as green building, are now working effectively as a market solution.

Institutional arrangements are also of high importance. Normally, government-led situations are common. There should be correct mechanisms to link policy-making processes and the implementation of work. The private sector should also lead, and in most cases support from civic groups are also important.

ON DESIGN ….

The practice of sustainable building is still at an early development stage. Standardisation processes are underway to help make our future buildings more adaptive and resilient to climate change or climate extreme. Design and construction of buildings play a key role. Life-cycle consideration is being promoted in the construction industry throughout the region. Many design tools and standards are in place, which speed up the process of sustainable construction.

The advance in technology in various aspects has made our buildings capable of withstanding extreme conditions resulting from climate change. High-performance buildings and innovations are now becoming mainstream in the market for wider adoptions. These movements are driving the practitioners for even better buildings.

The ultimate goal of the industry is to de-carbonise our development, that is, de-couple our urbanisation process and urban living with carbon emissions. Zero carbon building is not beyond reach and will be prevailing in the coming years. Yet, the real challenge lies within the infrastructure for energy. Micro energy grid and district energy systems are being promoted in the eco-cities development in the region. Their success will not only resolve the carbon emission issues but also make the infrastructure itself more resilient to climate extreme.

ON PEOPLE ....

In Asia, the development of high-rises and high-density built environment is a matter of necessity rather than choice, as rapid urbanisation has created huge amounts of pressure on housing and land. Yet, the experience in Asia has demonstrated that high-rise compact cities are not necessarily equal to inferior built environments and living conditions. The solution is hinged on the mind-set of practitioners on planning and building design. The experience of Hong Kong has demonstrated how to take the opportunity of the SARS pandemic and start a new planning process that is inductive to better living environments for both outdoor and indoor conditions.

Sustainable building designs and operation frameworks have been widely adopted worldwide, and most locations have their own specific systems tailor-made for their particular locations. However, building-oriented framework has the limitation in the control of the public realm and spaces beyond one's own building, and it is difficult to make up a neighbourhood or community. Hence, people have been working on the sustainability framework for a community to better cover the site scale aspects including social wellbeing, transport, neighbourhood design, and economy, and so on.

The change in behaviour needs design deliberation to create "change Agents". In Hong Kong, Taiwan, Korea, and Japan, cases were identified to influence the configuration of built environment design and operation.

WAY FORWARD ....

The process of urbanisation will be intensified in the region as regional countries are currently in the process of developing their economy. A virtuous cycle that is hinged on policy, design, and people regarding climate extreme is essential. The region has amassed a great deal of experience and viable solutions. Regional countries should take advantage of regional economic integration and build the capacity for addressing the issues of sustainable development.

References

[1] United Nations Development Programme (2015). Human Development Report Work for Human Development. UNDP.

[2] Clifford ML (2015). The greening of Asia: The business case for solving Asia's environmental emergency. US: Columbia University Press.

[3] ESCAP, KOICA (2012). Low Carbon Green Growth Roadmap for Asia and the Pacific: Turning Resource Constraints and the Climate Crisis into Economic Growth Opportunities. Bangkok: United Nations Publication.

[4] United Nations Division for Sustainable Development (1992). Agenda 21, Brazil: United Nations Publication.

[5] Robertson M (2014). Sustainability principles and practice: Principles and practice. UK. Routledge.

[6] US Department of Energy (2012). Climate Change and Infrastructure, Urban Systems, and Vulnerabilities. Technical Report for the US Department of Energy in Support of the National Climate Assessment.

[7] Baeumler A, Ijjasz-Vasquez E, Mehndiratta S (2012). Sustainable Low-Carbon City Development in China. Washington DC: The World Bank.

[8] Baeumler A, Ijjasz-Vasquez E, Mehndiratta S (2012). Sustainable Low-Carbon City Development in China. Washington DC: The World Bank.

[9] Newman P, Matan A (2013). Green urbanism in Asia: The emerging green tigers. Singapore: World Scientific Publishing Co Pte.

[10] Montgomery C (2014). Happy city: Transforming our lives through urban design. US: Anchor Canada.

[11] Robertson M (2014). Sustainability principles and practice: Principles and practice. UK: Routledge.

[12] Hitchcock D, Willard M (2009). The business guide to sustainability: Practical strategies and tools for organizations (2nd Ed.). London: Earthscan Publications.

[13] Robertson M (2014). Sustainability principles and practice: Principles and practice. UK: Routledge.

[14] Chan E, Qian Q, Lam P (2009). The market for green building in developed Asian cities—the perspectives of building designers. Energy Policy, 37(8), 3061–3070. doi:10.1016/j.enpol.2009.03.057

Building Sustainability in East Asia: Policy, Design, and People, First Edition. Vincent S Cheng and Jimmy C Tong.
© 2017 John Wiley & Sons Ltd. Published 2017 by John Wiley & Sons Ltd.

[15]   ESCAP, KOICA (2012). Low carbon green growth roadmap for Asia and the Pacific: Turning resource constraints and the climate crisis into economic growth opportunities. Bangkok: United Nations Publication.

[16]   BCA (2009). Singapore Leading the way for Green Building in the Tropics. Building and Construction Authority Singapore.

[17]   AIJ, IBEC (2005). Architecture for a sustainable future: All about the holistic approach in Japan. (A.I. (AIJ), Ed.) Tokyo, Japan: Institute for Building Environment and Energy Conservation (IBEC).

[18]   BCA (2009). Singapore Leading the way for Green Building in the Tropics. Building and Construction Authority Singapore.

[19]   Robertson M (2014). Sustainability Principles and Practice. UK: Routledge.

[20]   McGee T (2001). Urbanization Takes on New Dimensions in Asia's Population Giants. Population Reference Bureau.

[21]   Roberts B, Kanaley T (2006). Urbanization and Sustainability in Asia - Case Studies of Good Practice. Asian Development Bank.

[22]   ESCAP (2010). Urbanisation Trends in Asia and the Pacific [Fact sheet]. Retrieved from http://www.unescapsdd.org/files/documents/SPPS-Factsheet-urbanization-v5.pdf

[23]   Gretchen L (2016). Asia – Pacific Human Development Report - Shaping the Future: How Changing Demographics can Power Human Development, UNDP.

[24]   ESCAP (2010). Urbanisation Trends in Asia and the Pacific [Fact sheet]. Retrieved from http://www.unescapsdd.org/files/documents/SPPS-Factsheet-urbanization-v5.pdf

[25]   JMIAC, SB (2016). Population and Households. Japan Statistical Yearbook (Chapter 2). Retrieved from http://www.stat.go.jp/english/data/nenkan/pdf/yhyou02.pdf

[26]   Gretchen L (2016). Asia – Pacific Human Development Report - Shaping the Future: How Changing Demographics can Power Human Development, UNDP.

[27]   Chen M, Zhang H, Liu W, Zhang W (2014). The Global Pattern of Urbanization and Economic Growth: Evidence from the Last Three Decades. PLoS ONE 9(8): e103799. doi:10.1371/journal.pone.0103799

[28]   Chen M, Zhang H, Liu W, Zhang W (2014). The Global Pattern of Urbanization and Economic Growth: Evidence from the Last Three Decades. PLoS ONE 9(8): e103799. doi:10.1371/journal.pone.0103799

[29]   FocusEconomics (2016). China Economy - GDP, inflation, CPI and interest rate. Retrieved June 30, 2016, from http://www.focus-economics.com/countries/china

[30]   The World Bank (2009). World Development Report 2009 – Reshaping Economic Geography. Washington DC: World Bank.

[31]   International Energy Agency (2009). World Energy Outlook 2009, International Energy Agency.

[32]   Xiong, Z, Freney J, Mosier A, Zhu Z, Lee Y, Yagi K (2008). Impacts of Population Growth, Changing Food Preferences and Agricultural Practices on the Nitrogen Cycle in East Asia. Nutr Cycl Agroecosyst. 80: 189–198. doi:10.1007/s10705-007-9132-4

[33]   Roberts B, Kanaley T (2006). Urbanization and Sustainability in Asia - Case Studies of Good Practice. Asian Development Bank Report.

[34]   United Nations (2014). World Urbanization Prospects: The 2014 Revision. Retrieved from https://esa.un.org/unpd/wup/Publications/Files/WUP2014-Highlights.pdf

[35]   Gretchen L (2016). Asia – Pacific Human Development Report - Shaping the Future: How Changing Demographics can Power Human Development, UNDP.

[36]   ESCAP (2010). Urbanisation Trends in Asia and the Pacific [Fact sheet]. Retrieved from http://www.unescapsdd.org/files/documents/SPPS-Factsheet-urbanization-v5.pdf

[37]   Hein J, Preston F (2014). Asia's Sustainable Development A Literature Survey, Fung Global Institute.

[38]   US Department of Energy (2012). Climate Change and Infrastructure, Urban Systems, and Vulnerabilities, Technical Report for the US Department of Energy in Support of the National Climate Assessment. Retrieved from http://www.esd.ornl. gov/eess/Infrastructure.pdf

[39]   Chow A (2016). The Benefit of Incorporating Climate Change Impacts into Coastal Structures Design. Proceeding at ICE HKA Annual Conference 2016: Sustainable and Resilient Coastal Management Mitigating and Adapting to Climate Change. ICE, HK.

[40]   Information Office of the State Council of the People's Republic of China (IOSCPRC) (2011). China's Policies and Actions for Addressing Climate Change. Retrieved from http://english.gov.cn/archive/white_paper/2014/09/09/content_281474986284685. htm

[41]   Tracy A, Trumbull K, Loh C (2006). The Impacts of Climate Change in Hong Kong and the Pearl River Delta. Civic Exchange, Hong Kong SAR, China. Retrieved from http://www.hkccf.org/download/iccc2007/31May/S6B/Christine%20LOH/ The%20Impacts%20of%20Climate%20Change%20on%20Hong%20Kong%20and %20the%20Pearl%20River%20Delta.pdf

[42]   Levermore GJ (2008). A Review of the IPCC Assessment Report Four, Part 1: The IPCC Process and Greenhouse Gas Emission Trends from Buildings Worldwide, Building Services Engineering Research and Technology, 29(4), 349–361.

[43]   Yao R, Li B, Steemers K, (2005). Energy Policy and Standard for Built Environment in China. Renew Energy 30(13), 1973–1988.

[44]   Walsh K (2012). Accelerating Green Building in China, CIERP Policy Brief, Energy, Climate, and Innovation Program. The Fletcher School. Retrieved from http:// fletcher.tufts.edu/~/media/Fletcher/Microsites/CIERP/Publications/2012/ CIERPpolicyBrief_Walsh.pdf

[45]   Berkelmans L, Wang H (2012). Chinese Urban Residential Construction, Bulletin, Reserve bank of Australia, Retrieved from http://www.rba.gov.au/publications/ bulletin/2012/sep/pdf/bu-0912-3.pdf

[46]   >HKSAR Environment Bureau (2016). Greenhouse Gas Emissions in Hong Kong by Sector, Retrieved from http://www.epd.gov.hk/epd/sites/default/files/epd/english/ climate_change/files/HKGHG_Sectors_201606.pdf

[47]   Earley R, Kang L, An F, Green-Weiskel L (2011). Electric Vehicles in the Context of Sustainable Development in China. United Nations Department Of Economic And Social Affairs.

[48]   US Energy Information Administration (2016). International Energy Outlook 2016, US Energy Information Administration, Retrieved from http://www.eia.gov/ forecasts/ieo/pdf/0484(2016).pdf

[49]   He J (2015). China's INDC and Non-Fossil Energy Development, Advances in Climate Change Research, 6(3-4), 210–215.

[50]   Shin S (2015). Four point plan for promoting renewable energy in South Korea | Center for Energy and Environmental Policy. Retrieved June 14, 2016, from http:// ceep.udel.edu/four-point-plan-for-promoting-renewable-energy-in-south-korea/

[51]   Zhang D, Huang G, Xu Y, Gong Q (2015). Waste-to-Energy in China: Key Challenges and Opportunities. Energies, 8(12), 14182–14196. doi:10.3390/en81212422

[52]   Ryu C, Shin D (2013). Combined Heat and Power from Municipal Solid Waste: Current Status and Issues in South Korea. Energies, 6(1), 45–57. doi:10.3390/ en6010045

[53] Asia Business Council (2010). Containing Pandemic and Epidemic Diseases in Asia. Retrieved from http://www.asiabusinesscouncil.org/docs/DiseaseBriefing.pdf

[54] Robertson M (2014). Sustainability Principles and Practice. UK: Routledge.

[55] HC (2012). The Sustainable Development Timeline - 2012. Retrieved June 14, 2016, from http://www.iisd.org/library/sustainable-development-timeline-2012

[56] C40, Arup (2015). Climate Action in Megacities: C40 Cities Baseline and Opportunities. Retrieved from http://www.cam3.c40.org/images/C40ClimateActionInMegacities3.pdf

[57] World Economic Forum (2014). Future of Urban Development Initiative: Tianjin Champion City Strategy. Retrieved from http://www3.weforum.org/docs/WEF_IU_FutureUrbanDevelopmentInitiative_Tianjin_2013.pdf

[58] World Economic Forum (2014). Future of Urban Development Initiative: Dalian and Zhangjiakou Champion City Strategy. Retrieved from http://www3.weforum.org/docs/WEF_IU_FutureUrbanDevelopment_DalianZhangjiakou_ChampionCity Strategy_2014.pdf

[59] C40. Retrieved June 14, 2016, from http://www.c40.org/

[60] C40. Retrieved June 14, 2016, from http://www.c40.org/

[61] Sally R (2010). Regional Economic Integration in Asia: the Track Record and Prospects. ECIP Occasional Paper No.2/2010.

[62] ESCAP, UNEP, UNU, IGES (2016). Transformation for Sustainable Development - Promoting Environmental Sustainability in Asia and the Pacific. United Nations.

[63] PADECO (2010). Cities and Climate Change Mitigation: Case Study on Tokyo's Emissions Trading System, World Bank.

[64] Robertson M (2014). Sustainability Principles and Practice. UK: Routledge.

[65] The Climate Change Act and UK regulations. Retrieved June 14, 2016, from https://www.theccc.org.uk/tackling-climate-change/the-legal-landscape/global-action-on-climate-change/

[66] Delia R (2005). Regulatory Impact Analysis in OECD Countries Challenges for Developing. OECD.

[67] Jacob K, Weiland S, Ferretti J, Wascher D, Chodorowska D (2011). Integrating the Environment in Regulatory Impact Assessments. OECD.

[68] Delia R (2005). Regulatory Impact Analysis in OECD Countries Challenges for Developing. OECD.

[69] OECD (2010). Regulatory Policy and the Road to Sustainable Growth. Retrieved from https://www.oecd.org/regreform/policyconference/46270065.pdf

[70] ESCAP (2010). Urbanisation Trends in Asia and the Pacific [Fact sheet]. Retrieved from http://www.unescapsdd.org/files/documents/SPPS-Factsheet-urbanization-v5.pdf

[71] ESCAP, UNEP, UNU, IGES (2016). Transformation for Sustainable Development - Promoting Environmental Sustainability in Asia and the Pacific. United Nations.

[72] ESCAP, UN-Habitat (2015). The State of Asian and Pacific Cities 2015 Urban Transformations - Shifting from Quantity to Quality. Retrieved from http://www.unescap.org/sites/default/files/The%20State%20of%20Asian%20and%20Pacific%20Cities %202015.pdf

[73] ESCAP (2010). Urbanisation Trends in Asia and the Pacific [Fact sheet]. Retrieved from http://www.unescapsdd.org/files/documents/SPPS-Factsheet-urbanization-v5.pdf

[74] Athukorala PC (2010). ADB Working Paper Series on Regional Economic Integration- Production Networks and Trade Patterns in East Asia: Regionalization or Globalization? Asian Development Bank.

[75] Ren J, Du J (2012). Evolution of Energy Conservation Policies and Tools: The Case of Japan. Energy Procedia, 17, 171–177. doi:10.1016/j.egypro.2012.02.079

[76]   Mathews JA (2012). Green growth strategies—Korean initiatives. Futures, 44(8), 761–769. doi:10.1016/j.futures.2012.06.002

[77]   Mathews JA (2012). Green growth strategies—Korean initiatives. Futures, 44(8), 761–769. doi:10.1016/j.futures.2012.06.002

[78]   Kang SI, Kim H, Oh J (2012). Korea's Low-Carbon Green Growth Strategy. OECD Development Centre Working Papers. doi:10.1787/5k9cvqmvszbr-en

[79]   Rauch JN, Chi YF (2010). The Plight of Green GDP in China. Journal of Sustainable Develop-ment 3(1), 102–116.

[80]   Baeumler A, Ijjasz-Vasquez E, Mehndiratta S (2012). Sustainable Low-Carbon City Development in China. The World Bank.

[81]   Colletta N, Lim T, Kelles-Viitanen A (2001). Social Cohesion and Conflict Prevention in Asia – Managing Diversity Through Development. The World Bank.

[82]   Ministry of the Environment and Water Resources, Ministry of National Development (2014). Sustainable Singapore Blueprint 2015. Retrieved from http://www.mewr.gov.sg/ssb/files/ssb2015.pdf

[83]   Commission of the European Communities (2009). Mainstreaming Sustainable Development into EU Policies: 2009 Review of the European Union Strategy for Sustainable Development. Retrieved from http://eur-lex.europa.eu/LexUriServ/LexUriServ.do?uri=COM:2009:0400:FIN:en:PDF

[84]   Commission of the European Communities (2015). Sustainable Development in the European Union - 2015 Monitoring Report of the EU Sustainable Development Strategy. Retrieved from http://ec.europa.eu/eurostat/documents/3217494/6975281/KS-GT-15-001-EN-N.pdf

[85]   ESCAP, UNEP, UNU, IGES (2016). Transformation for Sustainable Development - Promoting Environmental Sustainability in Asia and the Pacific. United Nations.

[86]   Shen L, He B, Jiao L, Zhang X (2015). Research on the Development of Main Policy Instruments for Improving Building Energy-Efficiency, J. Cleaner Production 112(2), 1789–1803. doi:10.1016/j.jclepro.2015.06.108

[87]   Jang E, Park M, Roh T, Han K (2015). Policy instruments for Eco-Innovation in Asian Countries. Sustainability, 7(9), 12586–12614. doi:10.3390/su70912586

[88]   Liu R, Xu Y (2014). Comparison of International Incentive Policy of Green Building. Proceedings of the 5th International Asia Conference on Industrial Engineering and Management Innovation. doi:10.2991/iemi-14.2014.58

[89]   Khanna N, Romankiewicz J, Feng W, Zhou N, Ye Q (2014). Comparative Policy Study for Green Buildings in US and China. Lawrence Berkeley National Laboratory Report. doi:10.2172/1134231

[90]   Lee F (August 14, 2013). Hong Kong Pays a High Price for Its Low Water Fees. South China Morning Post. Retrieved June14, 2016, from http://www.scmp.com/comment/insight-opinion/article/1296528/hong-kong-pays-high-price-its-low-water-fees

[91]   Clark H (2013). The Importance of Governance for Sustainable Development, Remarks by Helen Clark, UNDP Administrator, on the occasion of the Singapore Lecture Series. Retrieved June 14, 2016, from http://www.undp.org/content/undp/en/home/presscenter/speeches/2012/03/13/the-importance-of-governance-for-sustainable-development.html

[92]   Li Y, Chen P, Chew D, Teo C (2014). Exploration of Critical Resources and Capabilities of Design Firms for Delivering Green Building Projects: Empirical Studies in Singapore. Habitat International, 41, 229–235.

[93]   IEEP, IES, ICF GHK, Naider (2015). Study to Analyse Differences in Costs of Implementing EU Policy. European Commission.

[94]   Buildings Department, Hong Kong SAR Government (2015), Practice Note for Authorized Persons, Registered Structural Engineers and Registered Geotechnical Engineers APP-130.

[95]   Planning Department, Hong Kong SAR Government (2011), Urban Climatic Map and Standards for Wind Environment – Feasibility Study – Stakeholders Engagement Digest.

[96]   Harris P (2012). Environmental Policy and Sustainable Development in China – Hong Kong in global Context. Bristol: Policy Press.

[97]   DJSI Family Overview. Retrieved from http://www.sustainability-indices.com/index-family-overview/djsi-family-overview/index.jsp

[98]   UNEP, DTIE, ETB (2006). Ways to Increase the Effectiveness of Capacity Building for Sustainable Development. UNEP.

[99]   Li Y, Chen P, Chew D, Teo C (2014). Exploration of Critical Resources and Capabilities of Design Firms for Delivering Green Building Projects: Empirical Studies in Singapore. Habitat International, 41, 229–235.

[100] Dissemination of CASBEE. Retrieved June 14, 2016, from http://www.ibec.or.jp/CASBEE/english/statistics.htm

[101] Ahn Y, Jung C, Suh M, Jeon M (2016). Integrated Construction Process for Green Building. Procedia Engineering, 145, 670–676. doi:10.1016/j.proeng.2016.04.065

[102] Li Y, Chen P, Chew D, Teo C (2014). Exploration of Critical Resources and Capabilities of Design Firms for Delivering Green Building Projects: Empirical Studies in Singapore. Habitat International, 41, 229–235.

[103] Larsson N (2009). The Integrated Design Process; History and Analysis. Retrieved June 14, 2016, from http://www.iisbe.org/node/88

[104] Ahn Y, Jung C, Suh M, Jeon M (2016). Integrated Construction Process for Green Building. Procedia Engineering, 145, 670–676. doi:10.1016/j.proeng.2016.04.065

[105] WBDG Sustainable Committee (2015). Sustainable. Retrieved June 14, 2016, from https://www.wbdg.org/design/sustainable.php

[106] Faiz A, Sumiani Y, Noorsaidi M (2013). A Review of the Application of LCA for Sustainable Buildings in Asia. AMR Advanced Materials Research, 724-725, 1597–1601. doi:10.4028/www.scientific.net/amr.724-725.1597

[107] Simiu E, Scallan R (1986). Wind effects on structures. New York: Wiley.

[108] American Society of Civil Engineers (2013). ASCE7: Minimum Design Loads for Bldgs and Other Structure. Retrieved from https://law.resource.org/pub/us/cfr/ibr/003/asce.7.2002.pdf

[109] Li C, Hong T (2014). Revisit of Energy Use and Technologies of High Performance Buildings. Environmental Energy Technologies Division.

[110] Brager G, Arens E (2014). Creating High Performance Buildings: Lower Energy, Better Com-fort. Presented at 4th International Congress in Advances in Applied Physics and Materials Science (APMAS 2014). Berkeley, California.

[111] Day J, Gunderson D (2014). Understanding High Performance Buildings: The Link between Occupant Knowledge of Passive Design Systems, Corresponding Behaviors, Occupant Comfort and Environmental Satisfaction, Building and Environment, 84, 114–124. doi:10.1016/j.buildenv.2014.11.003

[112] Integrated Environment Solutions. Integrated Analysis tools. Retrieved June 14, 2016, from http://www.iesve.com/software/ve-for-engineers

[113] Hall IJ, Prairie RR, Anderson HE, Boes EC (1978). Generation of a Typical Meteorological Year. Proceedings of the 1978 Annual Meeting of the American Section of the International Solar Energy Society. Denver, CO, 669–671.

[114]   Chan ALS, Chow TT, Fong SKF, Lin JZ (2006). Generation of a typical meteorological year for Hong Kong. Energy Conversion and Management, 47(1), 87–96. doi:10.1016/j.enconman.2005.02.010

[115]   Planning Department, Development Bureau, HKSAR Government (2007). Hong Kong 2030 Planning Vision and Strategy. Retrieved from http://www.pland.gov.hk/pland_en/p_study/comp_s/hk2030/eng/finalreport/pdf/E_FR.pdf

[116]   Development Bureau, HKSAR Government (2010). Optimising Land Use. Retrieved June 14, 2016, from http://www.devb.gov.hk/industrialbuildings/eng/background/optimising_land_use/index.html

[117]   Wan KSY, Yik, FWH (2004). Building design and energy end-use characteristics of high-rise residential buildings in Hong Kong. Applied Energy, 78(1), 19–36. doi:10.1016/s0306-2619(03)00103-x

[118]   Electrical & Mechanical Services Department, HKSAR Government (2013). Energy Con-sumption Indicators and Online Benchmarking Tools. Retrieved June 14, 2016, from http://ecib.emsd.gov.hk/en/index.htm

[119]   Chung W, Hui YV (2009). A study of energy efficiency of private office buildings in Hong Kong. Energy and Buildings, 41(6), 696–701. doi:10.1016/j.enbuild.2009.02.001

[120]   Shiming D, Burnett J (2002). Energy use and management in hotels in Hong Kong. International Journal of Hospitality Management, 21(4), 371–380. doi:10.1016/s0278-4319(02)00016-6

[121]   Electrical & Mechanical Services Department, HKSAR Government (2013). Energy Con-sumption Indicators and Online Benchmarking Tools. Retrieved June 14, 2016, from http://ecib.emsd.gov.hk/en/index.htm

[122]   Hui S (2015). Critical Evaluation of Zero Carbon Buildings in High Density Urban Cities. ZCB Journal, 3, 16–23.

[123]   Hui SCM (2010). Zero Energy and Zero Carbon Buildings: Myths and Facts. Proceeding of the International Conference on Intelligent Systems, Structures and Facilities (ISSF2010), 15–25.

[124]   Kang J, Lim J, Choi G, Lee S (2013). Building Policies for Energy Efficiency and the Development of a Zero-Energy Building Envelopment System in Korea. AMR Advanced Materials Research, 689, 35–38. doi:10.4028/www.scientific.net/amr.689.35

[125]   Kibert C (2012). The Emerging Future of Sustainable Construction: Net Zero, SHB2012, Proceeding of 7th international Symposium on Sustainable Healthy Buildings, 269–282.

[126]   Song Y, Sun J, Li J, Xie D (2014). Towards Net Zero Energy Building: Collaboration-based Sustainable Design and Practice of the Beijing Waterfowl Pavilion. Energy Procedia, 57, 1773–1782. doi:10.1016/j.egypro.2014.10.166

[127]   Zhao X, Pan W (2015). Delivering Zero Carbon Buildings: The Role of Innovative Business Models, Procedia Engineering, 118, 404 – 411.

[128]   Eurpoean Commission. Nearly zero-energy buildings. Retrieved July 22, 2016, from https://ec.europa.eu/energy/en/topics/energy-efficiency/buildings/nearly-zero-energy-buildings

[129]   California Public Utilities Commission (2015). New Residential Zero Net Energy Action Plan 2015-2020, California Energy Commission Efficiency Division and California Public Utilities Commission Energy Division.

[130]   Ng TSK, Yau RMH, Lam TNT, Cheng VSY (2013). Design and commission a zero-carbon building for hot and humid climate. International Journal of Low-Carbon Technologies, 11(2), 222–234. doi:10.1093/ijlct/ctt067

[131]   Building and Construction Authority, Singapore. The Island's First Retrofitted Zero Energy Building. Retrieved June 14, 2016, from www.bca.gov.sg/zeb/default/html

[132]   Liu H, Qiu G, Shao Y, Daminabo F, Riffat SB (2010). Preliminary experimental investigations of a biomass-fired micro-scale CHP with organic Rankine cycle. International Journal of Low-Carbon Technologies, 5(2), 81–87. doi:10.1093/ijlct/ctq005

[133]   Cheng V, Tong J (2014). Effect of Urban Density on PV Performance. In Gill MA (Ed.). Photovoltaics: Synthesis, Applications and Emerging Technologies (173–196), New York: Nova Science Publishers, Inc.

[134]   Tong J (2014). Numerical Simulations of Small Wind Turbines - HAWT Style. In Abraham JP and Plourde B (Eds.). Small-Scale Wind Power: Design, Analysis, and Environmental Impacts (129–146). New York: Momentum Press.

[135]   Kaushal J, Basak P (2014). Microgrid: An Emerging Power System and its Deployment in Indian Power Scenario. Paper presented at TEQIP Sponsored National Conference on Integrated Computational Techniques in Electrical Engineering (ICTEE-2014), Thapar University, Patiala, Punjab, India.

[136]   Marnay C, Zhou N (2008). Status of Overseas Microgrid Programs: Microgrid Research Activities in the US, Ernest Orlando Lawrence Berkeley National Laboratory Report, LBNL-56E. US: Environmental Energy Technologies Division.

[137]   Quiggin D, Cornell S, Tierney M, Buswell R (2012). A simulation and optimisation study: Towards a decentralised microgrid, using real world fluctuation data. Energy, 41(1), 549–559. doi:10.1016/j.energy.2012.02.007

[138]   UNEP (2014). District Energy in Cities - Unlocking the Full Potential of Energy Efficiency and Renewable Energy. United Nations Environment Program.

[139]   Hopkins E, Ferris J (2015). Place-Based Initiatives in the Context of Public Policy and Markets: Moving to Higher Ground. Sol Price School of Public Policy, University of Southern California.

[140]   Armbrister DM (2015). The Importance of Resident Engagement in Place-Based Initiatives, In Hopkins EM and Ferris JM (Eds.). Place-Based Initiatives in the Context of Public Policy and Markets: Moving to Higher Ground (59-62). Sol Price School of Public Policy, University of Southern California.

[141]   Chen L, Ng E (2012). Outdoor thermal comfort and outdoor activities: A review of research in the past decade. Cities, 29(2), 118–125. doi:10.1016/j.cities.2011.08.006

[142]   Du X, Bokel R, Dobbelseen A (2014). Building microclimate and summer thermal comfort in free-running buildings with diverse spaces: A Chinese vernacular house case. Building and Environment, 82, 215–227. doi:10.1016/j.buildenv.2014.08.022

[143]   Jamei E, Rajagopalan P, Seyedmahmoudian M, Jamei Y (2016). Review on the impact of urban geometry and pedestrian level greening on outdoor thermal comfort. Renewable and Sustainable Energy Reviews, 54, 1002–1017. doi:10.1016/j.rser.2015.10.104

[144]   Klemm W, Heusinkveld B, Lenzholzer S, Jacobs M, Hove B (2015). Psychological and physical impact of urban green spaces on outdoor thermal comfort during summertime in The Netherlands. Building and Environment, 83, 120–128. doi:10.1016/j.buildenv.2014.05.013

[145]   Yang W, Wong N, Jusuf S (2013). Thermal comfort in outdoor urban spaces in Singapore. Building and Environment, 59, 426–435. doi:10.1016/j.buildenv.2012.09.008

[146]   Zhou Z, Chen H, Deng Q, Mochida A (2013). A Field Study of Thermal Comfort in Outdoor and Semi-outdoor Environments in a Humid Subtropical Climate City. Journal of Asian Architecture and Building Engineering JJABE, 12(1), 73–79. doi:10.3130/jaabe.12.73

[147] Newman P, Matan A (2013). Green Urbanism in Asia: The Emerging Green Tigers. Singapore: World Scientific Publishing Co. Pte. Ltd.

[148] AB (2012). City rankings Hong Kong's best. The Economist. Retrieved June 14, 2016, from http://www.economist.com/blogs/gulliver/2012/07/city-rankings

[149] CLC, ULI (2013). 10 Principles for Liveable High Density Cities: Lessons from Singapore. Centre for Liveable Cities, Urban Land Institute.

[150] Yeun B, Yeh A (2011). High-rise Living in Asian cities, Springer. Chapter 1 High-rise living in Asian Cities. Netherlands: Springer.

[151] Bradecki T (2009). Mapping Urban Open Space and The Compact City – Research Methodology, Real Corp 2009: Cities 3.0 – Smart, Sustainable, Integrative. Retrieved from http://www.academia.edu/11820408/Mapping_urban_open_space_and_the_compact_city_research_methodology

[152] HK Building Department (2010). Builds Department Environment Report 2010. Retrieved from http://www.bd.gov.hk/english/documents/COER2010_eng.pdf

[153] Planning Department of HKSAR, UHK, UP, HKUST, PAC, UV (2008). Urban Climatic Map and Standards for Wind Environment-Feasibility Study. School of Architecture, CUHK.

[154] Information Services Department (2015). Hong Kong: The Facts. Retrieved from July 22, 2016, http://www.gov.hk/en/about/abouthk/factsheets/docs/population.pdf

[155] Mirzaei, PA, Haghighat F (2010). Pollution removal effectiveness of the pedestrian ventilation system. Journal of Wind Engineering and Industrial Aerodynamics, 99(1), 46–58. doi:10.1016/j.jweia.2010.10.007

[156] Mei SL (2014). Inferring Air Pollution by Sniffing Social Media. Advances in Social Networks Analysis and Mining (ASONAM). Paper presented at 2014 IEEE/ACM International Conference (534-539). Beijing: IEEE.

[157] MoEP (2012). Technical regulation on ambient air quality index (on trial). Beijing: Ministry of Environmental Protection of the People's Republic of China.

[158] Lv Y, Xu C, Yan J, Xu M (2013). Recent development of Low Carbon Community: An Overview. Paper ID: ICAE2013-059. Paper presented at The 5th International Conference on Applied Energy, South Africa.

[159] What is Placemaking? - Project for Public Spaces. Retrieved June 14, 2016, from http://www.pps.org/reference/what_is_placemaking/

[160] Fung A (2011). Environmental Protection and Testing: The Housing Authority's Experience. Speech presented at Innovation and Technology commission's Networking Dinner.

[161] HDB. Public Housing – A Singapore Icon. Retrieved June 14, 2016, from http://www.hdb.gov.sg/cs/infoweb/about-us/our-role/public-housing--a-singapore-icon

[162] HDB (2015). HDB Annual Report 2014/2015: Key Statistics. Retrieved June 14, 2016, from http://www10.hdb.gov.sg/ebook/ar2015/key-statistics.html

[163] Meegan R, Mitchell A (2001). 'It's Not Community Round Here, It's Neighbourhood': Neighbourhood Change and Cohesion in Urban Regeneration Policies. Urban Studies, 38, 2167–2194.

[164] Davies W, Herbert D (1993). Communities within Cities: an Urban Social Geography, London: Belhaven Press.

[165] Wellman B (2001). Physical Place and Cyber-place: Changing Portals and the rise of net-worked individualism. Int. J. Urban Regional Res., 25(2), 227–252.

[166] GBCA (2012) Green Star Communities PILOT Submission Guideline, p.21

[167] GBCA (2012) Green Star Communities PILOT Submission Guideline, p.21

[168] BISS. Making Collaboration Happen: Scaling Up Business Impacts on Sustainable Living. Retrieved June 14, 2016, from http://www.scp-centre.org/our-work/biss/

[169]   Broer S, Titheridge H (2011). Enabling Low Carbon Living in New Housing Developments – A Triple Bottom Line Analysis. In: Sustainability in Energy and Buildings: Results of the Second International Conference on Sustainability in Energy and Buildings (SEB'10). (189–199). Berlin: Springer Berlin Heidelberg.

[170]   Cubukcu E (2013). Walking for Sustainable Living. Procedia - Social and Behavioral Sciences, 85, 33–42. doi:10.1016/j.sbspro.2013.08.335

[171]   Hussein H, Jamaludin A (2014). POE of Bioclimatic Design Building towards Promoting Sustainable Living. Procedia - Social and Behavioral Sciences, 168, 280–288. doi:10.1016/j.sbspro.2014.10.233

[172]   Tomitsch M (2014). Towards the Real-Time City: An Investigation of Public Displays for Be-havior Change and Sustainable Living. Proceeding of the 7th Making City Liveable Conference, Kingscliff (NSW), Australia.

[173]   The World Bank. CO2 emissions (metric tons per capita). Retrieved July 22, 2016, from http://data.worldbank.org/indicator/EN.ATM.CO2E.PC

[174]   International Energy Agency, Definition and Simulation of Occupant Behavior in Buildings, Retrieved July 22, 2016, from http://www.annex66.org/?q=Publication

[175]   Lopes A, Gill A, Fam D (2015). Design and Social Practice Theory: A Promising Dialogue for Sustainable Living, J. Design Research, 13(3), 237–247.

[176]   Tomitsch M (2014). Towards the Real-Time City: An Investigation of Public Displays for Be-havior Change and Sustainable Living. Proceeding of the 7th Making City Liveable Conference, Kingscliff (NSW), Australia.

[177]   Winter T (2013). An uncomfortable truth: Air-conditioning and sustainability in Asia. Environment and Planning A, 45(3), 517–531. doi:10.1068/a45128

Index

Building Sustainability in East Asia: Policy, Design, and People, First Edition. Vincent S Cheng and Jimmy C Tong.
© 2017 John Wiley & Sons Ltd. Published 2017 by John Wiley & Sons Ltd.

Printed and bound by CPI Group (UK) Ltd, Croydon, CR0 4YY

27/10/2024

14580358-0001